ENVIRONMENTAL HEALTH AND SAFETY AUDITS

ENVIRONMENTAL HEALTH AND SAFETY AUDITS

A COMPENDIUM OF THOUGHTS AND TRENDS

Lawrence B. Cahill

Lanham • Boulder • New York • London

Published by Bernan Press
An imprint of The Rowman & Littlefield Publishing Group, Inc.
4501 Forbes Boulevard, Suite 200, Lanham, Maryland 20706
www.rowman.com
800-865-3457; info@bernan.com

Unit A, Whitacre Mews, 26-34 Stannary Street, London SE11 4AB

ISBN: 978-1-59888-811-9
E-ISBN: 978-1-59888-812-6

∞™ The paper used in this publication meets the minimum requirements of American National Standard
for Information Sciences—Permanence of Paper for Printed Library Materials, ANSI/NISO Z39.48-1992.

Printed in the United States of America

To my wondrous sons, Brendon and Bryan.
One calls me dad, the other calls me pops.
Makes me feel like I'm twice the father I really am . . .

Contents

Acknowledgments

I have conducted hundreds of EH&S audits and led some 200 training workshops all over the world. In conducting these assignments I have had my fair share of 16-hour plane rides, lost hotel reservations and luggage, and dinners served with still-moving creatures. I don't think I could have kept my sanity without my fellow road warriors by my side. They were and are both work colleagues and clients but best of all friends who shared the journey. Thanks to all of you, especially Ray Kane, Lori Michelin, Vance Merolla, Rob Costello, Joe Baker, Bob Reich, Ed Mongan, Wes McNealy, and Darwin Wika.

Thanks also to Mike Bittner, the editor of the on-line *EHS Journal*. The *Journal* has given me a wonderful forum to communicate my ideas and thoughts about the EH&S auditing profession. Keep it going Mike; it's worth it.

1

Introduction

"I think the teaching profession contributes more to the future of our society than any other single profession."

—John Wooden

The environmental auditing profession had its beginnings in the mid-1970s when the Securities and Exchange Commission (SEC) undertook enforcement actions against three major U.S. public companies. The SEC determined that the three companies were inconsistent in the reporting of the extent of their environmental liabilities in the press ("substantial") versus shareholder annual reports ("inconsequential"). The companies were required to hire third parties to conduct facility environmental audits to reconcile the differences and to clearly define the companies' liabilities.

Since the SEC initiatives some 40 years ago, the profession of environmental (and health and safety) auditing has taken hold and evolved significantly. With this evolution, regulatory agencies and organizations have adopted auditing standards and guidelines along the way (see Table 1.1). All of these expectations have essentially built upon the very first effort put forward by the U.S. EPA in 1986. The EPA in its landmark 1986 Audit Policy[1] described the seven elements of an effective environmental audit program. These include:

- Explicit top management support for environmental auditing and commitment to follow-up on audit findings
- An environmental auditing function independent of audited activities
- Adequate team staffing and auditor training
- Explicit audit program objectives, scope, resources, and frequency
- A process of collecting, analyzing, interpreting, and documenting information sufficient to achieve audit objectives
- Specific procedures to promptly prepare candid, clear, and appropriate written reports on audit findings, corrective actions, and scheduling for implementation
- Quality assurance procedures to assure the accuracy and thoroughness of environmental audits

TABLE 1.1
Thirty Years of EH&S Auditing Standards & Guidelines

Organization	Year	Standard or Guideline
ISO	2011	Guidelines for Auditing Management Systems (19011:2011); Supersedes ISO 19011:2002
BEAC	2008	Performance and Program Standards for the Professional Practice of EH&S Auditing
ASTM	2006	Standard Practice for Environmental Regulatory Compliance Audits (E2107-06)
USSC	2004	U.S. Sentencing Guidelines, Chapter 8, Effective Compliance and Ethics Program
ISO	2002	Guidelines for Quality and/or Environmental Management Systems Auditing (19011:2002); Supersedes 14010, 14011, 14012
USEPA	2000	EPA Incentives for Self-Policing
TAR	1997	Standards for the Design and Implementation of an EH&S Audit Program
ISO	1996	General Principles of EMS Auditing; EMS Auditing Procedures; Qualification Criteria for Environmental Auditors (14010, 14011, 14012)
TAR	1993	Standards for the Performance of EH&S Audits
USEPA	1986	Environmental Auditing Policy Statement-Elements of Effective Environmental Auditing Programs

Explanation of Acronyms:

ASTM – ASTM International; formerly American Society for Testing and Materials
BEAC – Board of Environmental Health and Safety Auditor Certifications
ISO – International Organization for Standardization
TAR – The Auditing Roundtable
USEPA – U.S. Environmental Protection Agency
USSC – U.S. Sentencing Commission

These elements have general applicability to any EH&S audit program and have been incorporated into most audit programs along with subsequent standards developed by ASTM, BEAC, The Auditing Roundtable, ISO, and other organizations with standing and legitimacy in the EH&S auditing community.

While the standards have set the foundation for expectations, organizations such as the Auditing Roundtable have established venues and forums where important additional audit issues could be discussed and sorted through. The thoughts and ideas of those practicing the profession of EH&S auditing early on and in subsequent years have been documented in presentations, articles, and books[2] on EH&S auditing. Still, even 40 years after the first audits were conducted, issues remain unresolved. This book addresses many of those outstanding issues by attempting to answer the following questions:

- How do we best measure our audit program's success and effectiveness?
- How can we better design our audits so that we're sure that we're identifying the real risks posed by our sites and not just documenting administrative deficiencies?
- How can we improve our auditors' writing skills in this age of Twitter?
- How are international audits different from domestic audits?
- Where are our auditors likely to mess up, and how can we prevent this from happening?
- How do we know when our auditors identify repeat findings on subsequent audits at the same facility; findings which truly demonstrate a lack of operational commitment, a systemic breakdown, or both?

- What's the difference between environmental audits and health and safety audits, and how do we train our auditors to assure that they're capable of doing both?
- How do we schedule and prioritize our audits so that we're maximizing our effectiveness while conserving our limited resources appropriately?
- What are the costs and benefits of outsourcing an audit program to a third party; does this make sense?
- Is statistically representative sampling expected on our third-party audits, and, even if it isn't, should we adopt the practice?

The author of this book draws from 35 years of auditing experience in more than 25 countries to address these issues. Along the way he draws upon his personal experiences to add some realism and humor in order to better tell the story. Experienced auditors will no doubt be able to relate directly to the experiences, and new auditors can learn from them as they become more polished and capable audit professionals.

Notes

1. U.S Environmental Protection Agency, "Environmental Auditing Policy Statement," *Federal Register*, Vol. 51, No. 131, pp. 25004–25010.

2. For example: Cahill, L.B., *Environmental Health and Safety Audits*, Government Institutes, Inc. Nine Editions, 1983–2011.

2

U.S. EH&S Regulatory Trends in 2015

How They Affect EH&S Audit Programs

"The care of human life and happiness, and not their destruction, is the first and only object of good government."

—Thomas Jefferson

Overview

Achieving full compliance with U.S. environmental health and safety (EH&S) regulations has always been challenging, but never more so than now. There have been some very interesting trends over recent years as demonstrated in this chapter.[1] Highlights include:

- The total pages of codified federal U.S. EH&S regulations now approach 30,000, with environmental regulations representing 88% of the total. Clean Air Act regulations alone account for 65% of the total number of environmental regulations.
- USEPA's budget has declined almost 25% in the past five years, yet its enforcement budget alone remains 40% larger than OSHA's entire budget.
- Industry-friendly programs such as EPA's Performance Track Program and Audit Policy and OSHA's Voluntary Protection Program have been de-emphasized or eliminated. At the same time, required release data provided to the EPA by industry under the Community Right-to-Know Act, makes the impact of industrial operations more transparent to the general public.
- EPA issued $163 million in civil and criminal penalties in FY2014, which may not seem like much compared to the $5.6 billion issued in FY2013. However, the great majority of the penalties issued in that year were due to the 2010 Deepwater Horizon incident. Still, by comparison, the $163 million represents almost 30% of OSHA's annual budget.
- The financial consequences of non-compliance globally have become more substantial. The stock prices of BP and the Tokyo Electric Power Company have still not recovered from notorious incidents in 2010 and 2011, respectively.

In sum, the challenges are still present for the regulated community as regulations continue to be burdensome and the consequences of non-compliance severe. With the 2016 presidential election and Republicans controlling both houses of Congress in 2015, the future of EH&S regulations remains mostly a mystery.

A Continuing Growth in U.S. EH&S Federal Regulations

Each year, the U.S. Government publishes the Code of Federal Regulations (CFRs), effective on July 1st of that year. The individual volumes for each year are usually available in January or February of the following year. The EPA is responsible for Title 40 of that Code, and OSHA is responsible for Title 29. In early 2015, the EPA and OSHA released the 2014 CFRs effective July 1, 2014. The total page count for 2014 was higher than 29,000 pages, with EPA regulations accounting for 88% of the total (see Figure 2.1). For the EPA this represented a one-year decrease of 125 pages (0.05%). Interestingly, it was the first time in history that there was an actual decrease in the number of pages, modest as it was. On the other hand, for OSHA the change amounted to a 13% increase in the number of pages compared to 2013.

It's important to note that approximately 65% of the pages in Title 40 of the CFR are devoted to Clean Air Act regulations (see Figure 2.2). This represents roughly 16,800 pages, meaning

FIGURE 2.1
Growth of EPA and OSHA Regulations

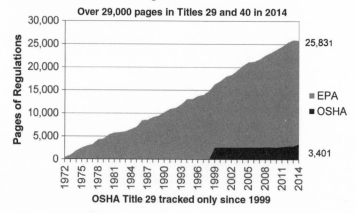

Source: U.S. Code of Federal Regulations

FIGURE 2.2
Distribution of EPA's Regulation by Media

Clean Air Regulations Represent 65% of the Total

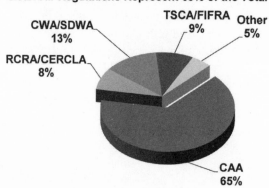

Source: U.S. Code of Federal Regulations – July 1, 2014

that the Clean Air Act alone has five times as many pages of regulations as the entire Title 29 CFR. Water regulations, governed by the Clean Water Act and the Safe Drinking Water Act, account for only 13% of Title 40. The bottom line is that federal air regulations are massive and in many cases complex and confusing. This is a case where more is not necessarily better.

What Might the Future Hold?—Changes in the Semi-Annual Regulatory Agenda

One could ask the question—Will the total page count ever decrease substantially? That's hard to say although this is possible where duplicative or outdated regulations are eliminated. As an example, on November 21, 2014, EPA and OSHA issued their Fall 2014 Semi-Annual Regulatory Agendas (as part of the Executive Branch's Unified Agenda) in the Federal Register, which is a spring and fall requirement for all regulatory agencies.[2] The EPA listed 186 additional regulations (not pages, but regulations) that either had been recently promulgated (but not yet codified) or were under development. OSHA listed only 33 regulations in its agenda, so the EPA will likely keep its substantial lead in regulatory page count (see Figure 2.3).

Eventually, all of these new regulations will be added to Titles 29 and 40, suggesting that there is no end in sight. It is very interesting to note that each of EPA's Semi-Annual Regulatory Agendas from 1999 to 2011 listed somewhere between 350 and 450 regulations under development, a consistency that is stunning. This means that for over a decade, there were always around 400 regulations under development on the docket. Interestingly, there has been a dramatic reduction in the number of regulations listed to roughly 200 from years 2012 to 2014. Maybe the regulatory pipeline is beginning to empty.

It is notable that on January 18, 2011 President Obama issued Executive Order 13563, "Improving Regulation and Regulatory Review." The Order required a government-wide review of existing rules "to remove outdated regulations that stifle job creation and make our economy less competitive. It's a review that will help bring order to regulations that have become a patchwork of overlapping rules."[3] At long last, there reportedly was a government-wide review of existing regulations to determine those that should be eliminated. The recent

FIGURE 2.3
USEPA/OSHA Regulatory Agenda Trends

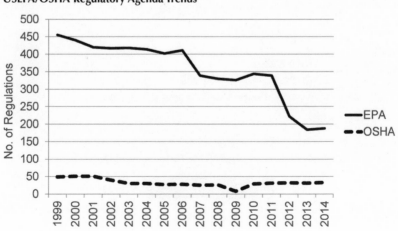

Source: U.S. Federal Register

U.S. EPA Regulatory Agendas and the low growth of regulations in the most recent Title 40 CFR seem to confirm that the Executive Order might be having the intended impact.

U.S. EPA and OSHA Budgets—A Comparison

From 2010 to 2015, the EPA's budget declined by 23%, from $10.3 billion to $7.9 billion (see Figure 2.4). What is even more interesting, however, is comparing OSHA and EPA budgets. As shown in Figure 2.5, EPA's budget is 14 times that of OSHA's. And even more startling is that EPA's enforcement budget alone is 40% larger than OSHA's entire budget!

A Shift in U.S. EPA's Audit Policy

Both the U.S. EPA and OSHA enacted Self-Disclosure Audit Policies in 2000.[4] These policies encourage the regulated community to conduct self-audits and report any regulatory

FIGURE 2.4
EPA's Annual Budget—A Downward Trend

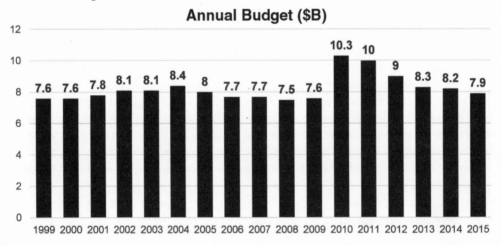

Source: U.S. EPA

FIGURE 2.5
Comparing EPA/OSHA Budgets

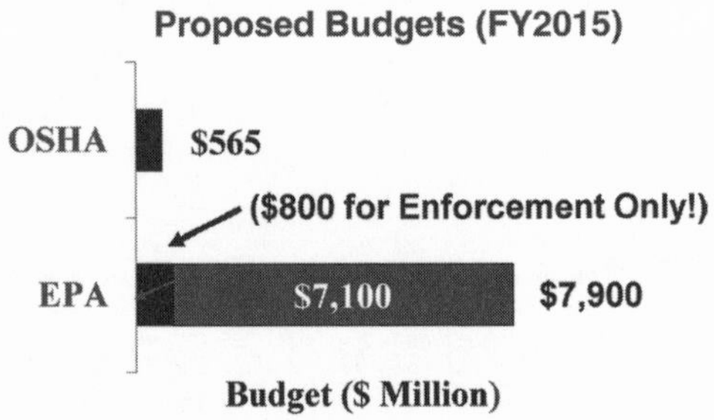

Source: U.S. EPA and OSHA

non-compliances to the respective agency. In turn, each agency will likely wave punitive penalties if certain criteria are met. And historic evidence has shown that in >90% of the EPA disclosure cases, no penalties were incurred.[5] For EPA in particular, use of the policy has been significant; disclosures have been resolved at over 13,000 facilities between FY1999 and FY2011.[6] Yet, the extensive use of the policy has led to a significant backlog of unresolved disclosures. This is a concern for the Agency and the regulated community. EPA has also become concerned that the vast majority of the disclosures have been simply deficiencies related to administrative and/or reporting requirements, particularly related to the routine reports required by the Emergency Planning and Community Right-to-Know Act (EPCRA). And finally, EPA's data show that only 0.35% of the pollutant reductions achieved by Agency actions have resulted from audit policy disclosures.

The EPA has recognized this as an issue with respect to the audit policy. According to the following language provided in the EPA's FY 2013 OECA National Program Manager Guidance, EPA's approach with respect to the Audit Policy is evolving. "Since implementation of the Audit Policy began in 1995, EPA's enforcement program has increased its understanding of environmental compliance auditing, and believes that internal reviews of compliance have become more widely adopted by the regulated community, as part of good management. In addition, EPA has found that most violations disclosed under the Policy are not in the highest priority enforcement areas for protecting human health and the environment. EPA believes it can reduce investment in the program to a limited national presence without undermining the incentives for regulated entities to do internal compliance reviews to find and correct violations. As we reduce investment in this program, EPA is considering several options, including a modified Audit Policy program that is self-implementing."[7]

In that regard, in June of 2015 EPA's Office of Enforcement and Compliance Assurance announced that it would be implementing an e-Disclosure portal in the fall of 2015. The portal will be a new, centralized next generation Web-based system for more efficiently receiving and processing violations disclosed to EPA under its self-disclosure policies. There will be a two-tiered disclosure hierarchy. For Tier 1 routine disclosures, the portal "will automatically issue an electronic Notice of Determination confirming that the violations are resolved with no assessment of civil penalties conditioned on the accuracy and completeness of the submitter's certified eDisclosure. Non-routine Tier 2 disclosures will be handled pretty much the same as they have been historically."[8]

The EPA losing interest in industry-friendly programs is not without precedent. For example, the National Environmental Performance Track ("Performance Track") program was a public-private partnership that encouraged continuous environmental improvement through environmental management systems, community outreach, and measurable results. The program identified "stars" among the regulated community and these facilities received certain benefits (e.g., fast track permitting) as a result of their outstanding environmental performance. Performance Track was terminated by EPA on May 14, 2009. This cancellation of the program was one of the very first initiatives under the first Obama administration. At the time of termination, the program had a membership base of 547 facilities in 49 states and Puerto Rico. A major part of the rationale for the elimination of the program was that the "star" participants were still experiencing significant non-compliance issues. That is, there didn't appear to be a solid correlation between participation and regulatory performance.

The Department of Labor's Office of Inspector General found similar issues with the VPP Star Program in 2013. According to the OIG report[9] OSHA did not have sufficient controls to ensure VPP worksites maintained exemplary occupational safety and health systems:

According to the OIG report: "Thirteen percent of participants had injury and illness rates above industry averages or were cited with violations of safety and health standards, but most of these participants were allowed to remain in the program. Moreover, OSHA policy allowed participants with injury and illness rates above industry averages to potentially remain in the program for up to 6 years, raising serious questions as to whether the companies were fully protecting their workers. The OIG made recommendations to the Assistant Secretary for Occupational Safety and Health covering policies, controls, and oversight so OSHA can better ensure only VPP participants with exemplary safety and health systems remain in the program."

U.S. EPA's Toxic Release Inventory

The Emergency Planning and Community Right-to-Know Act (EPCRA) Section 313 requires EPA, states and tribes to collect data annually on releases and transfers of certain toxic chemicals from industrial facilities and make the data available to the public in the Toxics Release Inventory (TRI). The current TRI toxic chemical list contains 594 individually-listed chemicals and 31 chemical categories (including four categories containing 68 specifically-listed chemicals). The data are extremely easy to extract and analyze by the general public. To the point where individual facilities can be assessed for releases on a county-by-county basis. This has made the environmental impacts of facility operations amazingly transparent. A map showing the 2013 TRI ranking of individual states is shown in Figure 2.6.

FIGURE 2.6
2013 TRI Total Release Rankings by State

Note: Number 1 refers to the state with the most toxic releases.
Source: Data from USEPA

U.S. EPA Enforcement Trends

In late 2014, the EPA released its enforcement results for fiscal year 2014, which ended on September 30, 2014. Included in those results were data on civil and criminal penalties assessed (see Figure 2.7). As shown, EPA issued $163 million in civil and criminal penalties in FY2014, which may not seem like much compared to the $5.6 billion issued in FY2013. However, the great majority of the penalties issued in that year were due to 2010 the Deepwater Horizon incident. Still, by comparison the $163 million represents almost 30% of OSHA's annual budget.

It's difficult to draw a definitive conclusion from the 2014 results. Still, with almost a billion dollars being spent by EPA on enforcement, it remains a serious issue, unmatched by any other country. As EPA has stated in its FY2013 OECA National Program Manager Guidance, we will "aggressively go after pollution problems that make a difference in communities. EPA will use vigorous civil and criminal enforcement that targets the most serious water, air and chemical hazards, as well as advance environmental justice by protecting vulnerable communities."[10]

A Retrospective Look at the Financial Impacts of Recent Major Incidents

History has shown that as far back as the Union Carbide Bhopal, India incident in 1984 EH&S incidents can have a major impact in the financial wellbeing of corporations. In fact, Union Carbide no longer exists as an entity; its assets purchased by Dow Chemical. Ironically, Dow's headquarters in Midland, Michigan, are picketed each year in March by people still protesting the Bhopal incident.

Two more recent incidents were the 2010 BP Deepwater Horizon Gulf of Mexico explosion and oil spill and the 2011 Japanese tsunami that took down a Tokyo Electric Power Company nuclear power plant.

FIGURE 2.7
Trends in USEPA Penalties

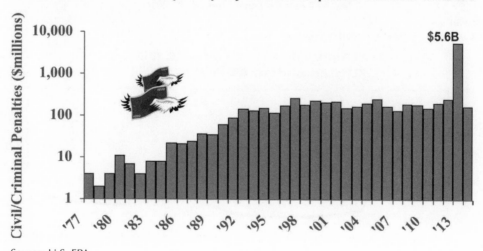

Source: U.S. EPA

BP's Deepwater Horizon incident did indeed have significant organizational and financial consequences on the company. In addition to having set up a $20 billion reimbursement fund for damages, the company's stock price dropped significantly, from over $60 per share to the low $40 per share (see Figure 2.8). Some five years later, on April 17, 2015 the company's stock price still hovered in the low $40 per share. BP was also removed from the Dow Jones Sustainability Index (DJSI) on May 31, 2010. Oh, and the company's Chief Executive Officer was forced to resign on October 1, 2010.

TEPCO is the owner of the Fukushima Dai-Ichi nuclear power plant where the power supply was knocked out by a March 2011 tsunami disrupting the cooling process that led to reactor meltdown. There were certainly very tragic human consequences resulting from this incident. However, the financial impacts on TEPCO were similarly severe (see Figure 2.9). On April 20, 2015, a full four years after the incident, the company's stock price was still 80% below where it was just before the incident. In addition, TEPCO was removed from the Dow Jones Sustainability Index (DJSI).

FIGURE 2.8
BP April 2010 GoM Platform Explosion and Spill

FIGURE 2.9
Tokyo Electric Power-March 2011 Tsunami

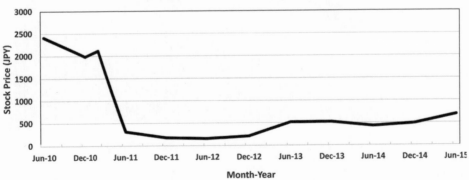

What conclusions can one draw from these experiences? Well, incidents can happen anywhere, anytime and typically nothing good will come from one. And second, a major loss in market capitalization, as occurred with BP and TEPCO, can also have significant human consequences. It puts employees' retirements at risk, it adversely affects investors, and it can hinder the ability to aid those who were directly impacted by the event.

Notes

1. This topic has been addressed previously by Mr. Cahill in the following articles: Cahill, L. B. (Spring 2013) Risk and Regulation: A Post-Election Update on Environmental, Health, and Safety Trends in the United States, *Environmental Quality Management*, John Wiley & Sons, Inc., New York, NY, 22(3); Cahill, L.B. (2011, Summer) Risk and Regulation: An Update on Environmental, Health, and Safety Trends in the United States. *Environmental Quality Management*. John Wiley & Sons, Inc., New York, NY, 20(4), and Cahill, L.B. (Autumn 2010) Achieving Environmental, Health, and Safety Compliance in the United States: Hard and Getting Harder. *Environmental Quality Management*. John Wiley & Sons, Inc., New York, NY 20(1). This chapter presents new data and information not previously published in the above articles.

2. The Regulatory Flexibility Act and Executive Order 12866 require Spring and Fall Regulatory Agendas.

3. Obama, Barak, "Toward a 21st-Century Regulatory System," *Wall Street Journal*, January 18, 2011.

4. U.S. EPA's Incentives for Self-Policing, Discovery, Disclosure, Correction and Prevention of Violations, Federal Register Vol. 65, No. 70, April 11, 2000, pp. 19618–19627; OSHA's Final Policy Concerning the Occupational Safety and Health Administration's Treatment of Voluntary Employer Safety and Health Self-Audits, Federal Register Vol. 65, No. 146, July 28, 2000, pp. 46498–46503.

5. Adam Kushner, Former Director of EPA's Office of Civil Enforcement and Partner, Hogan Lovells, "EPA's Implementation of Self-Disclosure Policy," Auditing Roundtable Presentation, September 6, 2012.

6. U.S. Environmental Protection Agency, Office of Enforcement and Compliance Assurance, National Enforcement Trends Report, June 2012, p. 126.

7. U.S. Environmental Protection Agency, FY2013 Office of Enforcement and Compliance Assurance (OECA), National Program Manager (NPM) Guidance, April 30, 2012, p. 15.

8. U.S. Environmental Protection Agency, Office of Enforcement and Compliance Assurance, eDisclosure Information Sheet, EPA 300-B-15-002, June 2015.

9. U.S. Department of Labor, Office of Inspector General, Voluntary Protection Program: Controls Are Not Sufficient to Ensure Only Worksites with Exemplary Safety And Health Systems Remain in the Program, December 16, 2013.

10. U.S. Environmental Protection Agency, FY2013 Office of Enforcement and Compliance Assurance (OECA), National Program Manager (NPM) Guidance, April 30, 2012, p. 6.

3

Measuring the Success of an EH&S Audit Program

What Are the Best Metrics?

"Success depends upon previous preparation, and without such preparation there is sure to be failure."

—Confucius

Introduction

Environmental, health and safety (EH&S) audit program managers often ask, "How do I know if my program is working?" This is certainly a legitimate question and is often a "pass-down" of the same question that is asked of the program managers by senior management, including the board of directors. This chapter explores the possible metrics that could be used to determine success or failure.[1] As with most things in life, the answer is not obvious and can be quite complex. Let's take a look at six common metrics that are often touted as valid measuring sticks:

- A reduction in environmental releases and workplace injuries
- Improved compliance as defined by a reduction in fines and enforcement actions
- A reduction in the total number of audit findings
- A reduction in the number of high-risk audit findings
- A reduction in the number of repeat audit findings
- A high rate of on-time closure of audit action items.

Analysis of these six metrics shows that none of them by itself provides a logically compelling performance measure. Of the group, the most compelling metric seems to be the last, a high rate of on-time closure of audit action items.

External Measures

The first two measures relate to how improved performance might be evaluated using external metrics: a reduction in incidents and a reduction in notable regulatory non-compliances.

1. A Reduction in Environmental Releases and Workplace Injuries

If only there were a direct correlation between the rigor of an audit program and a reduction in environmental releases and workplace injuries. But alas, there may be a relationship, but little evidence suggests a direct causality.

Certainly an audit program can contribute to improved performance but history shows that audit programs, no matter how rigorous, are not an adequate substitute for establishing sound management systems and controls at sites. Audit programs are meant to be verification programs; that is, the objective is to periodically verify compliance with applicable rules and regulations and self-imposed corporate standards and controls. Once audit programs are used as a surrogate for sound on-site EH&S management systems, the site is doomed to fail. Accountability must reside with site management and the systems and controls that have been implemented on site and not be reliant upon a check-up visit once every two to three years to "get back on track."

For any number of reasons (e.g., independent audits occur once every two to three years and are typically only a snapshot evaluation of compliance) expecting a reduction in releases and workplace injuries as a result of periodic audits misplaces the emphasis of who in fact is responsible. Sadly, there are too many cases where the audit program manager has become the "fall guy" when an incident takes place ("Why didn't the audit program identify the situation that resulted in the incident?").

Bottom line evaluation of this metric: Poor

2. Improved Compliance as Defined by a Reduction in Fines and Enforcement Actions

This measure is often considered by management to be a good way to determine the value of an audit program. However, the measure is typically relevant only in the United States where regulatory agency fines and enforcement actions are common. For example, the U.S. Environmental Protection Agency (USEPA) issued over $186 million in civil and criminal enforcement penalties in its last full fiscal year, and the U.S. Safety and Occupational Health Administration (OSHA) issued its largest fine ever of $87 million in 2010. Most other countries' regulatory agencies have a more cooperative approach with the regulated community; fines and enforcement actions are rare. So for multi-national companies this is not a very good measure.

Secondly, even in the United States the enforcement posture of the federal government can change from one presidential administration to the next. For example, under the Obama administration USEPA's budget has increased by 35%. In fiscal year 2011, the budget for the USEPA's Office of Enforcement and Compliance Assurance (OECA) alone is $618 million, the largest budget in OECA history and larger than OSHA's total budget. OECA states one of its main enforcement goals going forward as follows:

> Aggressively go after pollution problems that make a difference in communities. Vigorous civil and criminal enforcement that targets the most serious water, air, and chemical hazards . . .

Similarly, on June 18, 2010, OSHA published and made effective its Severe Violator Enforcement Program (SVEP) Directive. OSHA announced that it was "implementing the program to focus on employers who continuously disregard their legal obligations to protect their workers." As a result, the playing field has changed substantially over the past year and a half in the United States, and the number of fines and enforcement actions likely will rise, independent of the rigor of any company's audit program.

Bottom line evaluation of this metric: Poor

Internal Measures

The following four measures relate to how improved performance might be evaluated using internal metrics: a reduction over time in the number of total, high priority, or repeat findings and a high rate of closure of audit action items.

1. A Reduction in the Total Number of Audit Findings

This metric is rather easy to calculate but even easier to dismiss as not meaningful; principally because "all findings are not created equal." For example, let's say that an audit team finds that a regulatory program does not exist at a site because the site erroneously believes that the program does not apply to them. One finding. A second team visits the site two years later and finds that much work has been undertaken and the program has been largely implemented. However, there are still four administrative requirements that are not being met completely. Four findings. It's pretty clear that the one finding on the first audit far outstrips the importance of the four findings on the second audit. Hence, if one were using total number of findings as a measure, the results would be quite misleading.

Bottom line evaluation of this metric: Poor

2. A Reduction in the Number of High-Risk Audit Findings

Many companies rank individual audit findings by the level of risk posed to the organization. They might even use a scheme similar to the one provided below:

- **SIGNIFICANT: HIGHEST PRIORITY ACTION REQUIRED:** Situations that could result in substantial risk to the environment, the public, employees, stockholders, customers, the Company or its reputation, or in criminal or civil liability for knowing violations.
- **MAJOR: PRIORITY ACTION REQUIRED:** Does not meet the criteria for Level I but is more than an isolated or occasional situation. Should not continue beyond the short term.
- **MINOR: ACTION REQUIRED:** Findings may be administrative in nature or involve an isolated or occasional situation.

Thus, one potential metric would be the trend in the number of high-risk findings. Over time one would expect the number of high-risk findings to decrease as sites are audited a second and third time. Unfortunately, even with the definitions provided above there is often a lack of consistency in applying the ratings scheme, leaving some to question the merits of using this metric. Some of the reasons for inconsistency include:

- No matter how well vetted within the organization, the definitions leave room for interpretation.
- Some but not all auditors believe that no regulatory finding could possibly be minor.
- Some but not all auditors (and at times legal counsel) believe that all regulatory findings should be classified as significant.

In addition, many companies do not classify findings based on risk, believing that all findings are equally important. Obviously, for these companies, this metric is not appropriate.

Bottom line evaluation of this metric: Fair

3. A Reduction in the Number of Repeat Audit Findings

Many corporate audit programs are designed to capture and report on repeat findings on individual facility audits. A repeat finding can be defined as:

- A finding that had been identified in the previous independent audit of the same topic (e.g., environmental, employee safety) for which a corrective action has not been completed, or
- A finding of a substantially similar nature to one that had been identified, and reportedly corrected, in the previous independent audit of the same topic.

These repeat findings are typically considered serious findings and justifiably receive significant management attention.

The problem with using repeat findings as a valid metric for measuring performance is that most companies have not gone to the trouble of defining what is and what is not a repeat finding. This results in varying interpretations by auditors. One auditor might say that any exceedances of wastewater discharge limits would be a repeat finding had this been identified on the previous audit. Another auditor might look a bit deeper and determine that due to product changes the treatment plant might had to have been operated differently and that the current pH exceedances have a different root cause.

As a consequence, auditors really need to focus on the intent of the repeat findings classification before proceeding to label something as a "repeat." The question actually is: Did a breakdown in a management system really cause this repeat failure or was it simply an isolated case of a similar nature?

For example, on any given fire safety audit of a large manufacturing site, auditors, if they look long and hard enough, could almost always find an issue with inspections and maintenance of portable fire extinguishers. Should a missing inspection tag on one fire extinguisher out of a universe of several hundred constitute a repeat finding if another extinguisher was without a tag on the previous audit? Probably not, if the fire safety management system is found to be fundamentally sound. These situations should be thought of as recurring findings, not necessarily repeats. Similar situations can be found in other compliance areas where the universe of items to be audited is also quite large (e.g., MSDSs, hazardous waste manifests, wastewater discharges).

In sum, repeat findings should not be used as a performance metric unless everyone is working off the same playbook. Sites should not be punished by the repeat classification when the system is otherwise fully implemented and effective.

Bottom line evaluation of this metric: Good if the term "repeat finding" has been defined and a uniform understanding of the term is applied to the audit program.

4. A High Rate of On-Time Closure of Audit Action Items

Any audit results in a corrective action plan, usually developed by the management of the site that was audited. The plan includes: a description of the finding, the proposed corrective action, the person responsible, and the target date for completion. Many companies formally track the closure of these action items and calculate the percent of those that are completed on time. It's all about "Say what you do; and do what you say." This metric can be very useful in determining the value of and commitment to an audit program. Its benefits are:

- It's a simple measurement.
- The responsible individuals are the ones defining the actions and setting the dates.
- It's a true measure of management's commitment to compliance.
- Using a percent closure metric normalizes performance among differing operations.

Even with this metric, there are challenges. Some of those observed in companies that use this as a metric include:

- The numerator and denominator of the ratio (i.e., action items closed on time to total action items) should be clearly defined and reported consistently.
- Complete 100% on-time closure is unrealistic and should be challenged.
- Original timelines need a sanity check; there is a tendency to revise/extend dates when being tracked.
- Original target dates need to be fixed; unless an extension is reviewed and approved by a senior executive.

Let's take a brief look as to how this metric might be put into play. The following chart is an example of how action item closure might be presented for a company's five strategic business units (SBUs) for the latest six-month period.

FIGURE 3.1
Action Item Closure

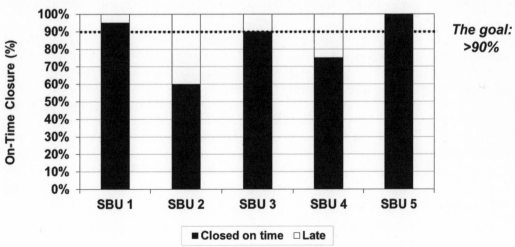

Where Are the Problems?

The goal is to have each SBU achieve a greater than 90% on-time closure rate, recognizing that the ultimate objective in a perfect world is 100%. What conclusions can be drawn from the results? First, SBUs 1 and 3 are in good shape and are meeting the goal. Second, SBUs 2 and 4 are not meeting the goal, with SBU2 achieving only a 60% on-time closure rate. This is clearly a red flag. And third, SBU5 should be praised for a perfect 100% or scrutinized for its perfection, depending upon the relative cynicism of the reviewer.

Bottom line evaluation of this metric: Good to Excellent

Overall Conclusion

When evaluating the effectiveness of an EH&S audit program, a third-party evaluator typically looks to answer three basic questions:

- Is there an accurate inventory of auditable sites and are these being audited at an acceptable frequency?
- Are the audits being conducted by knowledgeable and trained staff consistent with generally accepted audit practices?
- Are the findings being effectively communicated and closed consistent with the proposed action item closure dates?

This chapter has proposed that the closure of action items is key to a successful program. In an ideal world, a better metric for a successful program might indeed be fewer fines and incidents. But alas, we do not live in an ideal world.

Note

1. A version of this chapter was previously published in the on-line *EHS Journal* as: "Measuring the Success of an EHS Audit Program," August 23, 2010.

4

EH&S Audits

Have We Lost Our Way?

"To raise new questions, new possibilities, to regard old problems from a new angle, requires creative imagination and marks real advance in science."

—Albert Einstein

Background

The past few years have reminded us once again that mismanagement of environmental health and safety (EH&S) responsibilities can have a substantial impact on people, the environment, and a company's bottom line. What impact might these catastrophic events such as the Japanese tsunami and the BP Gulf of Mexico oil spill have on EH&S compliance and audit programs? Substantial, one would hope. Audit programs for decades have focused on achieving compliance with detailed administrative requirements. And, particularly in the U.S., this is not surprising. Currently, there are over 29,000 pages of EH&S regulations at the federal level (Titles 29 and 40 of the Code of Federal Regulations [CFR]) to say nothing of state and local requirements. And this number of CFR pages has grown by 3,000 in the past five years alone. So, not only are the requirements substantial, but they are changing and becoming more stringent all the time. It is no wonder that audit programs focus on these regulatory requirements, with penalties as high as $25,000 per day or more per violation per occurrence.

Although assessing compliance with detailed regulatory requirements such as inspections, permits, plans, manifests, material safety data sheets (MSDSs), reports, written procedures, and the like is important to achieving and maintaining compliance, recent events suggest that ignoring real EH&S risks can truly affect a company's bottom line. One would think that these potential outcomes should impact how audits are conducted now and in the future. In practice, this would beg the question, should one be more concerned about:

- A wastewater discharge that has had periodic, minor exceedances of pH or that the underground sewers are 50 years old and have never been surveyed with a camera to determine their integrity?
- The exact height of the containment wall of an above ground storage tank that is two-inches too short or that the tank has not been tested for integrity in the last 40 years?
- The expiration date of a single confined space entry permit that is not provided or that attendants at an entry are not always focused on the entry itself?
- The guarding on a seldom-used grinder in the maintenance shop that is not set at the precise gap defined by the regulations or that operators on the production line are routinely clearing debris without shutting down the line?

In each of the cases posed above, most traditional compliance audits focus on the former issue, not the latter; even though the latter in each case poses a higher risk. This is mostly because there are clear and precise requirements associated with the first, and not so much with the second.

It might be time to take a hard look at the objectives and philosophies of corporate audit programs. Achieving compliance with regulations is quite important, especially in the U.S., but identifying, assessing, and managing EH&S risks should also be at the core of any audit program. Maybe one way to do this is to not rely so much on detailed compliance audit protocols, containing thousands of questions, but to identify potential high-risk activities in each compliance area (e.g., air, water, confined space, lockout/tagout, etc.) that should be reviewed in detail on each and every audit; and to provide guidance on how that review should be conducted. For example, for a given audit, one would identify up front any high-risk activities (e.g., ammonia refrigeration, hazardous materials storage tanks and transfer systems, work areas where respirators are required, flammable storage buildings, high production areas requiring significant operator oversight, etc.) in need of special attention and then develop a specific audit plan on how to review that activity. This does not mean that the more mundane activities and operations are ignored. It simply means that there is a defined approach on how to audit individual high risk operations, within the context of the larger audit.

On a personal note, after reviewing hundreds of audit reports over the years, I am personally dismayed that we might have lost our way. When I review a finding that tells me that "one weekly inspection of a hazardous waste accumulation point was missed in the past six months" or that "two of two hundred fire extinguishers did not have updated monthly inspections," I truly wonder whether our audits are helping to protect people and the environment. This is not to say that a more risk-based audit approach would have prevented the Japanese tsunami or BP tragedies, but maybe it's worth a try. This chapter discusses one possible approach that could be taken to improve EH&S audits and make them more risk-based and valuable.[1]

Overview of the Methodology

The essence of the methodology is that each finding on an audit can and should be evaluated for the potential consequences of non-compliance with regulations or company standards *and* should also be evaluated for the inherent EH&S risk posed and the likelihood of an incident (see Figure 4.1). Findings with a low potential for non-compliance consequences and a low potential for EH&S risk and incidents (Quadrant I) are common, but their discovery probably does not add significant value to the audit or the organization. Findings with a high potential for non-compliance consequences and a high potential for EH&S risk and incidents

FIGURE 4.1
Prioritizing Finding on EH&S Audits

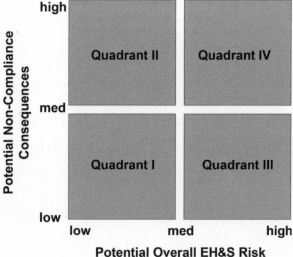

(Quadrant IV) are more rare but add significant value to the process and the understanding of the audited facility's operational status and inherent risk. Other findings might present either the high likelihood of regulatory compliance consequences or business or other risks but not both (Quadrants II and III) and should be evaluated accordingly.

One can use the outcomes of the methodology in a number of different ways. First, if 95% of an audit's findings are in Quadrant I, then either the site is performing remarkably well or the audit team did indeed focus principally on the detailed administrative requirements. If this result continues for facility after facility, in say a highly regulated and very risky business, then maybe the auditors need some additional training or the program needs to be re-worked.

A second way to utilize the outcomes is to look closely at the outliers. Obviously, the outliers in Quadrant IV are worthy of scrutiny. However, it's not so obvious that outliers in Quadrant III, in particular, warrant some attention. In fact, programs that focus only on regulatory compliance may miss Quadrant III findings entirely. These would be high-risk findings for which there is little in the way of a regulatory or corporate requirement driver. A good example, discussed later, would be a large water tank that is 50 years old and has never been tested for wall thickness or integrity. This oversight might be quite important as many water tanks at manufacturing facilities are located in the utilities area, often near an electrical substation. Should the tank rupture, the plant could very easily lose power. Not a good thing.

There are numerous other ways the outcomes could be used to benefit the organization. Only two are discussed above. In the examples provided later in this article, one can see a real-life application of the methodology and imagine other analyses that could be undertaken.

Examples of Findings for Each Quadrant

The question might be posed: Don't we need a definition of what findings go in what part of each quadrant? In a perfect world, that definition would be straightforward. Unfortunately, we do not live in a perfect world. The best way to provide guidance on quadrant placement is by way of examples. Provided below are examples based on the author's experiences. If an organization decides to utilize the methodology, the idea would be for the developer to provide additional examples based on a particular company's operations and compliance challenges. Definition by example is probably the best approach.

Quadrant I—Low Non-Compliance Consequences, Low Risk

- Minor exceedances of pH in wastewater discharge
- Secondary containment wall two inches too short and not proximate to sensitive receptors
- No expiration date on one confined space entry permit
- Guarding on seldom-used grinder in shop is not set at correct gap
- Community Right-to-Know Tier II inventory report submitted late

Quadrant II—Medium to High Non-Compliance Consequences, Low Risk

- No tracking of refrigerant leak rates and repairs for units containing >50 lbs of ozone depleting substances (ODS)
- Expired wastewater discharge permit; no re-application on record
- Missed Toxics Release Inventory (TRI) chemical on last five reports
- Incomplete hazardous waste manifests for wastes that are disposed properly
- Lack of documentation for hazard communication training that was conducted

Quadrant III—Low Non-Compliance Consequences, Medium to High Risk

- No knowledge of the integrity of 50-year-old process sewers
- Forty-year-old aboveground storage tank never tested for integrity
- Storage of incompatible materials together in containers and tanks
- Hydrogen storage trucks parked and unloaded onsite not exceeding Process Safety Management thresholds with no process safety controls
- Grandfathered landfill with no historical knowledge of what was disposed

Quadrant IV—High Non-Compliance Consequences, High Risk

- Attendants at active entry not always attentive or present
- Operators on production line clearing debris while line is active
- Large hazardous material tank on the banks of a high-quality designated stream with no secondary containment
- Unpermitted discharges of industrial wastewater
- Unreported and mitigated releases of hazardous substances to the environment

Creating a Hypothesis for Expected Outcomes

What might be an ideal distribution for findings using the proposed methodology? After some hypothesis testing and considerable discussions with auditors, it was suggested that the ideal distribution of audit findings might be a "leaf" distribution as shown in Figure 4.2. This theory

FIGURE 4.2
The Expected "Leaf" Distribution

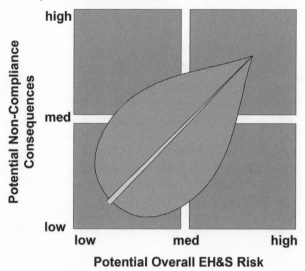

Potential Non-Compliance Consequences (y-axis: low, med, high)

Potential Overall EH&S Risk (x-axis: low, med, high)

would apply mainly to a mature audit program where sites have been through one or more audit cycles. Described below is the rationale for the theory. It might be an over-simplification, but typically simple is better.

First, it was suggested that there should not be a concentration of findings in the truly low/low category. This would imply that the audit team identified a significant number of minor, administrative deficiencies. Although they might indeed exist at a given site, an audit report is usually not the best place to record them. Many companies classify these as "local attention only" findings and do not include them in the audit report itself as they can create unnecessary "noise" when management reviews and acts upon the report findings. And that is why the leaf is not positioned at the low/low intersection but moved up and to the right.

Second, as one moves up both axes, one would expect to see a fanning out of the distribution in both directions. This would indicate the auditors are focusing on both regulatory compliance and relative risks.

And third, the distribution would hopefully taper off in the fourth quadrant (particularly for more mature audit programs) as it is unlikely that there would be many high risk, high compliance consequences findings. At least, one would hope not.

Using the Excel "Bubble Chart" Application

One could assume that simply plotting the points on a simple scattergram might obscure the fact that there were numerous findings with the same score. A better solution was to use the "bubble chart" application found in Excel. This better shows the findings distribution. And that is how the results are displayed for the case studies presented later in this chapter; but before presenting the results of four case studies, listed below is some guidance on interpreting the Excel-generated bubble charts:

- Each audit finding has been given a compliance score of 1–10 and a risk score of 1–10.
- The size of the bubble represents the number of findings with that particular score (e.g., 1,1; 4,3). The larger the bubble, the greater the number of findings.
- The white star represents the average score for all the findings.
- Scores ranging from 1–5 are considered to be in the lower quadrant; those ranging from 6–10 are considered to be in the higher quadrant.
- The auditors were given the following guidance for individual findings scores:

Compliance Scores:
- 1–3: Low likelihood of substantial regulatory agency fines or enforcement
- 4–6: Moderate likelihood of substantial regulatory fines or enforcement
- 7–9: Major likelihood of substantial regulatory fines or enforcement
- 10: Significant likelihood of substantial regulatory fines or enforcement

Risk Scores:
- 1–3: Low likelihood of incident occurrence; low severity or impact if incident occurs
- 4–6: Moderate likelihood of incident occurrence; moderate severity or impact if incident occurs
- 7–9: Major likelihood of incident occurrence; major severity or impact if incident occurs
- 10: Significant likelihood of incident occurrence; significant severity or impact if incident occurs

Four Case Studies

The methodology was tested on more than a dozen audits, both in the U.S. and in other countries, at a wide variety of sites using multiple, different audit teams. The results were quite interesting if not conclusive. The remainder of this chapter discusses some of the more interesting outcomes.

Case 1: A Safety and Health Audit in the U.S.

This audit had 49 findings (see Figure 4.3). The largest circle represents six findings with the same score and the smallest circles represent one finding with the same score. The average score was 4.2 for risk and 3.2 for compliance with 14% of the findings in the low/low classification (1,1 to 2,2).

In this audit of a major U.S. chemical plant, the resultant distribution is much like the leaf model. There are not too many low/low findings (14%), and the team appears to have focused on both regulatory compliance as well as risk, with the average risk score slightly higher than the average compliance score. All in all, it is an audit well done.

Case 2: A DOT Audit in Asia

This audit had 10 findings (see Figure 4.4). The largest circle represents three findings with the same score, and the smallest circles represent one finding with the same score. The average score was 3.8 for risk and 1.4 for compliance with 20% of the findings in the low/low classification (1,1 to 2,2).

FIGURE 4.3
S&OH Audit in the U.S.

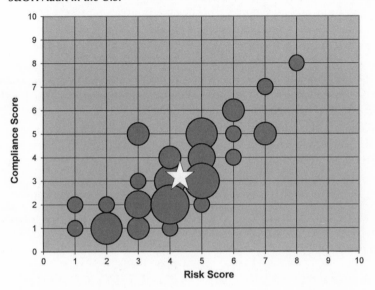

FIGURE 4.4
DOT Audit in Asia

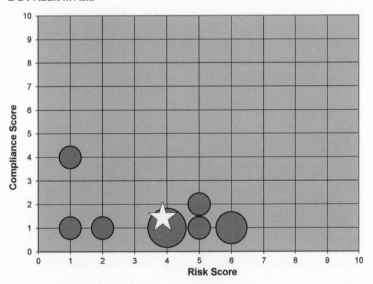

In this audit of a distribution center in Asia, there was only one finding with a compliance score above two, and 20% of the ten findings were in the low/low classification. It was not that surprising that there was only one finding in an upper quadrant given that it was a distribution center. But the lack of compliance findings was a bit surprising. The feedback given by the team was that there are very few if any distribution or transportation regulations in the country although this could not be confirmed. At least the analysis made us ask the question.

Case 3: An EH&S Audit in Europe

This audit had 49 findings (see Figure 4.5). The largest circles represent nine findings with the same score and the smallest circles represent one finding with the same score. The average score was 3.1 for risk and 1.4 for compliance, with 31% of the findings in the low/low classification (1,1 to 2,2).

This audit of a major agricultural operation proved to be very interesting. Note the very low average score for compliance (1.4) and that 31% of the findings were in the low/low classification. It turned out that the audit team members could neither speak nor read the local language and did no research in advance with respect to the country's EH&S regulations. Therefore, they focused on company standards and good management practices. The chart results made that approach pretty obvious. As a result, an action plan was put together for future audits to assure that the teams had at least one member who was fluent in the local language and that research was done in advance on applicable local regulations (i.e., purchase of a country-specific audit protocol).

Case 4: An EH&S Audit in the U.S.

This audit had 16 findings (see Figure 4.6). The largest circles represent two findings with the same score and the smallest circles represent one finding with the same score. The average score was 3.0 for risk and 2.2 for compliance, with 38% of the findings in the low/low classification (1,1 to 2,2).

This audit of a seasonal manufacturing facility resulted in a low compliance score (2.2) and 38% of the findings in the low/low classification (1,1 to 2,2). Here, it turned out that the processing area was not operating at the time of the audit; the plant was essentially shut down. Thus, the vast majority of findings were in Quadrant I. Historically, in this company these types of seasonal operations were mostly audited in the off-season; the rationale being that "we're just

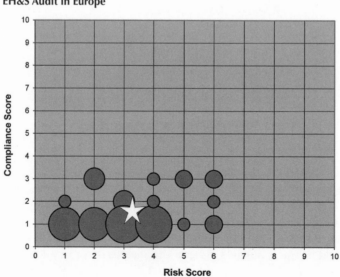

FIGURE 4.5
EH&S Audit in Europe

FIGURE 4.6
EH&S Audit in the U.S.

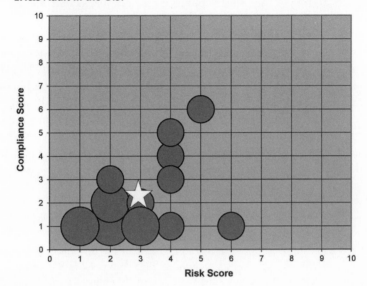

too busy to handle an EH&S audit during our busy season." That begs the question, of course: Wouldn't you definitely want to audit the site when activity is at its peak? In this case, the corrective action was to not audit these types of plants at the same time of year for each audit cycle.

Conclusion

Using a methodology similar to the one described in this chapter can help to improve the value of an EH&S audit program. For the cases discussed in this article, the analysis got to the root of some very important questions:

- Are we too heavily focused on minor administrative requirements?
- Do our auditors have the compliance tools and language skills to conduct audits in a variety of different countries?
- Are we visiting the sites at the best time of year to assure operations are active?

There are numerous other audit program issues that a findings risk analysis can uncover. Take a good, long, hard look at your audit program. Are you getting both what you want *and* what you need?

Some in the industry have identified the challenge discussed in this chapter as a critical audit program issue and developed new initiatives to improve their internal audit processes to better focus on risk. For example, David Cummings, DuPont's Corporate Process Safety Competency Leader, has undertaken a pilot program to improve the focus of PSM audits at his company. Starting in 2009, DuPont added mechanical integrity/quality assurance (MIQA) specialists to specific PSM audits at thirty higher hazard process sites in the U.S. This was initiated because MIQA is a critical component of an effective PSM program, and auditing MIQA is a niche specialty that in many cases requires significant practical experience, detailed technical knowledge, and an understanding of the proper application of not just OSHA requirements but also Recognized and Generally Accepted Good Engineering Practices (RAGAGEPs) developed by API, ASME, ISA, and others. Utilizing MIQA audit specialists also enables expanded data sampling and field evaluations as part of each review, which ultimately provides an improved risk-based assessment. The typical EH&S auditor does not have these skill sets. The DuPont MIQA auditing pilot program has proven to be very beneficial and is being applied in other regions.[2]

Finally, recent incidents suggest that a risk-based approach would add more value to audit programs. Historical and current audit findings can and should be evaluated for their value-added to the organization's understanding of EH&S risk. If that evaluation suggests that audit programs are not achieving the objective of better identifying and managing EH&S risks, then programs should be re-worked to accomplish that objective.

Notes

1. This chapter is based on a trilogy of articles published in the on-line *EHS Journal* by Lawrence Cahill and Robert J. Costello, PE, CPEA, Esq. including: "EHS Audits-Have We Lost Our Way, Part III," March 12, 2013; "EHS Audits-Have We Lost Our Way? A Sequel," August 13, 2011; and "EH&S Audits – Have We Lost Our Way?" July 11, 2010.

2. This information has been taken from discussions with Cummings and his presentation "Mechanical Integrity Challenges in PSM Audits – It's All About the Details," given at the Philadelphia National Meeting of The Auditing Roundtable, on September 9, 2010.

5

Where Have You Gone, Ernest Hemingway?

Tips on Writing Exceptional Audit Findings

"There is nothing to writing. All you do is sit down at a typewriter and bleed."

—Ernest Hemingway

Introduction

I've always thought that two of the most interesting challenges in one's career as an EH&S professional might be: (1) to write an article on effective writing and (2) to give a presentation on effective public speaking. Talk about putting yourself on the firing line. With this chapter, I will now have done both; so please be kind.[1]

A good percentage of my responsibilities these days is devoted to reviewing reports, procedures, plans, and various other EH&S compliance documents written by others. For the most part, this has been a very rewarding, but at times frustrating, experience. I worry that the emergence of PowerPoint, Twitter, and texting is beginning to affect the writing capabilities of younger professionals. The shorthand limit of 140 characters with Twitter, in particular, has at times resulted in people forgetting, or ignoring, what they learned about writing in high school; things like the different parts of speech and that every sentence must have a subject and verb and usually an object.

I recall that as an engineering major in college, I was required to take a public speaking course, but there was no available writing course directed at engineers. There was only freshman English, which consisted mostly of reading and discussing Dostoevsky and other famous writers, many of whom wrote their original manuscripts in their native languages. Was I really reading Dostoevsky or some unknown ghost writer who was paid to translate the classic *Crime and Punishment*? I wonder what is required of engineers and scientists in college today. I hope it's more than what was available to me. And I hope that the writing courses are not focused on how to write an effective PowerPoint bullet slide.

A Comparative Example

What has concerned me lately is what I have experienced in reviewing scores of findings in dozens of audit reports written by numerous auditors. Generally, findings are observations of a site's compliance deficiencies and, as such, must be both factual and understandable. And most importantly, when read, a finding should be crystal clear on how best to correct the deficiency. Alas, this is not always the case. Let's take the following example, which I recently came across in an audit report.

"Site not conducting weekly hazardous waste inspections, as required."

Did that finding meet my expectations? It certainly did not and when I did a spell check, Word did not find it acceptable either (fragment). For one, it is not crystal clear. And if one reads the statement literally, it actually says that the requirement is to "not" conduct weekly inspections. In terms of the parts of speech, the statement is missing an article ("the") and a verb, among other things. I've always believed that findings should be written in complete sentences so there is less room for misinterpretation.

So, let's see how the above finding could be improved and meet the stated criteria.

"The site could not demonstrate inspections were being conducted of the 90-day hazardous waste accumulation point located on the south side of Building A23. There were no completed inspections available for the previous six months. Federal hazardous waste regulations require the site to inspect 'areas where containers are stored, at least weekly, looking for leaks and for deterioration caused by corrosion and other factors.'"

Now, the issue is clear. The sentences are complete, and the evidence is presented. This approach might involve a bit more work, but the outcome is significantly better.

A Quick Exercise

I tend to get "preachy" about writing effective findings. You would too after reviewing hundreds of audit reports and thousands of findings. Presented below is a quick exercise using the findings discussed above. Match the word with the correct part of speech. (There may be more than one word identified with the same part of speech.) The answers are at the end of the chapter. Unlike most EH&S computer-based training, 70% is not a passing grade, but you can repeat the exercise until a 100% score is achieved!

Align the Word with the Part of Speech

Words	Parts of Speech
The	Adjective
Site	Adverb
Could not demonstrate	Article
Inspections	Conjunction
Of	Noun
Accumulation	Preposition
Point	Pronoun
And	Verb

Sorry to bring you back to your high school days. I hope they were pleasant memories.

Some Pet Peeves

Most, especially my high-school English teachers, would say that I am no grammarian, but I do have some clear opinions about what constitutes good writing. Many of these are strictly supported by grammarians; others less so. Still, right or wrong, these issues still drive me crazy as a reader and reviewer of audit reports. They are presented directly below.

Who vs. That

Many audit reports have statements much like: . . . the operator that was responsible for the inspections." In my opinion, the word "that" should be replaced by the word "who," although like most things in life, it's more complicated than that. According to the Grammarist website, "Most writers use *that* and *which* as the relative pronouns for inanimate objects and who as the relative pronoun for humans. This widespread habit has led to the mistaken belief that using *that* in reference to humans is an error. In fact, while most editors prefer who for people, there is no rule saying we can't use *that*, and *that* has been widely used in reference to people for many centuries."[2] To me it's "who," but there is some flexibility here.

The Word "Data"

According to Strunk and White: "Like strata, phenomena, and media, data is a plural noun and is best used with a plural verb. The word, however, is slowly gaining acceptance as a singular noun."[3] Here again, it's complicated. I prefer the plural, however, use of "data" as a singular noun is probably more common in practice.

Pompous Language and Hyperbole

Provided below are three phrases from actual audit reports:

- If this facility is allowed to operate as it has in the past, it will lead to a swath of destruction along the surrounding environment.
- There were a plethora of violations at the site.
- The chemical hygiene program was deficient and could be improved. This is a serious concern.

Words and phrases such as "plethora," "swath of destruction," "violations," and "serious concern" have no business appearing in an audit report. If one needs to communicate the seriousness of the finding, then use a significant, major or minor findings classification system, or its equivalent. And finally, follow Strunk and White's advice: "Do not be tempted by a twenty dollar word when there is a ten-center handy, ready, and able."[4]

Imprecision

There are many imprecise words that have no business appearing in an audit report. These include as examples "some, a few, several, insufficient, inadequate." For example, the following is not an acceptable audit finding: "Almost all of the employees working in Assembly Room B were not wearing the required hearing protection." If there are issues with hearing protection, the auditor should capture three numbers (how many employees there are, how many the auditor inspected, and how many were deficient). These numbers should be incorporated into the finding statement. Second, if something is "inadequate," the auditor must describe the specific inadequacy in the finding statement. Otherwise, a general statement of inadequacy does nothing to help the site or the reader correct the specific deficiencies.

A Negative, Reinforcing Phrase Following a Positive Phrase

Read the following statement carefully: "Three of the ten waste drums at the hazardous materials storage area had no labels, as required by the Federal Hazardous Waste regulations." A strict interpretation of this audit finding would imply that no labels are required on the drums. That is, remove the extraneous words and you have "waste drums had no labels, as required." The corrective action then is to remove the labels from the other drums with labels, which is not the outcome that anyone wants. The actual meaning of the finding is probably clear to most people. However, auditors must leave no room for doubt. In this case, the auditor would be best served by using two sentences, not one.

Ending a Sentence with a Preposition

It is common to see sentences in audit reports ending with "complied with," or "disposed of." This is the classic faux pas of ending a sentence with a preposition. It bothers me, but I have been told that it has become acceptable. Even Strunk and White says: "Years ago, students were warned not to end a sentence with a preposition; time of course, has softened that rigid decree."[5] Try to avoid the practice, but where it makes sense, it is fine.

Conjecture

Does the following statement bother you? "Because of the possibility of solvent releases into the sewers, the wastewater treatment ponds might have to be permitted as a hazardous waste surface impoundment." It should. This is classic conjecture. Conjecture is defined by Merriam-Webster as a conclusion deduced by guesswork. Findings in audit reports should be statements of fact, not guesses on the auditor's part.

Hearsay Evidence

Hearsay evidence should not be used in creating an audit finding. The two actual findings presented below are classic examples:

- A unit operator reported that occasionally attendants leave their station during confined space entries.
- The team was told that there have been several spills of hazardous materials by forklift operators.

The use of hearsay evidence is easily challenged by the site, as well as it should be. Like a good newspaper investigative reporter, there must be a second source before any conclusion is drawn from hearsay evidence.

The Word "Appears" and Its Variations

My very least favorite word that auditors occasionally use in audit reports is "appears" and its variations. Two examples are provided below:

- It appears that there is insufficient aisle space in the soap plant.
- Based on the lack of documented training records in their personnel files, it appears that Robert Atwell, David Lee Perry, and Rodrigo Munoz have not received the required annual Hazardous Waste training.

These statements suggest that the auditor was not interested enough in the process to actually determine whether the aisle space or training results were acceptable. Auditors need to work to the end; is the situation compliant or not? That's only fair to those being audited.

How to Get Better

"Make every word tell." —William Strunk, Jr., and E. B. White, *The Elements of Style*[6]

Improving your writing takes hard work and attention to detail. One of the very best resources to help in that process is the book *The Elements of Style* by Strunk and White. Yes, White is the E. B. White of *Charlotte's Web*. This book, originally crafted by Strunk when he was a Cornell University professor in the 1930s is now available in its fourth edition on Amazon.com or your local bookstore for less than $10. For me personally, it was a revelation; truly the best book on writing I've ever read. I have read it many, many times. The *New York Times* said in a book review: "Buy it. Study it. Enjoy it. It's as timeless as a book can be in our age of volubility."

I am routinely amazed that when I teach an EH&S auditing or compliance course and ask the students how many are aware of the "little book" as it is often called, that less than half the class raise their hands. What a shame. This book will teach you to write simply and effectively. Get it, read it, and apply the lessons. You won't be sorry.

Conclusion

"The most essential gift for a good writer is a built-in, shockproof crap detector. This is the writer's radar and all great writers have it." —Ernest Hemingway

Writing well is a challenge. However, it is a necessary skill for all EH&S professionals, especially auditors. Not having this arrow in your quiver can really hold you back. Don't let that happen.

Exercise answers: (1,c; 2,e; 3,h; 4,e; 5,f; 6,a; 7,e; 8,d).

Acknowledgment

Inspiration for the title of this article is from the 1967 movie *The Graduate* and Simon and Garfunkel's song "Mrs. Robinson," in which they ask, "Where have you gone, Joe DiMaggio; A nation turns its lonely eyes to you."

Notes

1. A version of this chapter was previously published by the author in the on-line *EHS Journal* as: "Where Have You Gone Ernest Hemingway?" November 20, 2013.
2. From grammarist.com/usage, "who versus that."
3. Strunk and White, Page 44.
4. Strunk and White, Page 76.
5. Strunk and White, Page 77.
6. Strunk Jr., William and E. B. White, *The Elements of Style,* Fourth Edition, Allyn & Bacon, 2000.

6

Twenty Lessons Learned on International Assignments

The Good and the Bad

"We shall not cease from exploration, and the end of all our exploring will be to arrive where we started and know the place for the first time."

—T. S. Eliot

Introduction

By my accounting, during my career I have been on more than 75 international trips and hundreds of U.S. trips. I've seen the Eiffel Tower, boated on the Amazon, seen the Reclining Buddha in Bangkok and the Little Mermaid in Copenhagen, taken in a play at the Sidney Opera House, and taught a course at a South African Big Game Reserve. On the other hand, I've also seen every gate at Philadelphia International Airport, been stranded in an Indianapolis hotel for three days during a snowstorm, been on a plane whose right engine caught fire during takeoff and was forced to return to the airport and make an emergency landing after dumping jet fuel all over South Philadelphia, spent 16 straight hours on a plane (two work days) several times, and arrived in Manila at midnight with no driver in sight. Typically, it's not the being there; it's the getting there that can break you.

A while ago I heard Terri Morrison, author of *Kiss, Bow or Shake Hands: The Bestselling Guide to Doing Business in More Than Sixty Countries* (Adams Media, 2006), captivate an audience for two hours on the tricks and traps of working internationally. It made me reflect on my personal travel experiences over the past three-plus decades, especially those overseas. Below are my top twenty experiences and the lessons I learned from each. I generally don't have that great of a memory, but these experiences are emblazoned in my brain as if each had happened yesterday. Each left a lasting impression.[1]

1. Arriving at Midnight in Cologne, Germany

When I was in my mid-20s and working for Exxon, I travelled to a petrochemical plant in rural France to conduct a month-long noise study. As I was finishing up, my boss sent me a cable (yes, a cable) requesting that I travel to Cologne, Germany, to assess the noise emissions of a new industrial furnace. And off I went on my first inter-country train ride. I arrived at

the Cologne train station at midnight expecting to be met by a fellow American who was working on a process unit startup at the plant. Sadly, I found out later that he was instructed to meet an earlier train, and when I was not on it he went back to his hotel. So, there I was at midnight having no idea what to do next. It turns out my instructions read that I should go to the "hotel." I picked up my luggage and gear and proceeded to walk around until I found the "hotel." As I began my walk, a completely inebriated fellow starts walking with me and chatting happily to me in German. I said back to him in English that I did not understand German, but it didn't matter to him in the slightest; he was having a grand old time with his new best friend. Finally, after about 45 minutes, with the time approaching 2:00 A.M., I saw a sign that said "hotel." Home at last, or so I thought. It turns out that it was not the hotel where I was registered. But miracle of miracles, the lad at the front desk called around to all the other hotels until he found the one where I was registered. It was within walking distance. Seeing how loaded down I was with luggage and equipment and not having been to Cologne before, the lad then had a bus boy load my stuff onto a hand truck and walk me over to my hotel. I will never forget the help I received that night and early morning. Oh, and my other friend finally got bored with me and went on his merry way.

Lesson Learned: Have a backup plan for your contingency plan.

2. Planes, Trains & Automobiles in Ecuador

Some years ago, I was asked to participate on an EH&S audit at an oil production field in the Ecuadorian rain forest along the Amazon River. It turned out to be quite an adventure in travel. It began with a plane ride from Philadelphia to Miami, where I boarded a second plane to Quito, the capital city of Ecuador, which is at the highest elevation above sea level of any capital in any country (9,350 feet). The next day was a flight of about two hours to a secondary city and then onto a boat for two hours down the Amazon. This was followed by a half-hour trip in the back of a pickup truck to the compound. As we arrived at the gate there were several guards with automatic weapons. It seems that five company employees had been kidnapped (and subsequently released) the week before by the indigenous Indians. There was a controversy (and there still is) about who owned the oil that was being extracted and how the Indians should be compensated. Needless to say, the audit team was a bit on edge during that evaluation.

Lesson Learned: International business travel is more exhausting and nerve wracking than glamorous.

3. Driving Around in South Korea

In auditing several sites in South Korea, I was being driven by the client's South Korean audit program manager. He is a great guy and to this day I consider him a friend. But during one conversation in the car I mentioned that I had recently worked with one of his company colleagues who was Japanese and I thought very competent. My friend for some reason immediately became distant for the next half day. I couldn't figure it out until I realized that Japan had occupied Korea during World War II and that bad feelings still existed. Now, that was over a half century ago, so you might wonder why the resentment still existed. Well, we only have to look at the U.S. and realize that some still hold resentment over the outcome of the Civil War, which ended in 1865. Go figure.

Lesson Learned: Learn about the history of the country you're visiting.

4. Security in Cainelands, South Africa

At this major manufacturing site, we initially met with the plant manager in his office. He had a beautiful, air-conditioned office with a wonderful view, except that he kept his blinds closed. This seemed rather odd, and we asked why. He did not answer directly but took us outside the building to show us the bullet holes in the exterior wall outside his office. Occasionally local bandits armed with automatic weapons would rob the payroll truck when it arrived with the cash for the salaries of the employees. Corrective action – They no longer pay the employees in cash, although the plant manager quite rationally remains skittish.

Lesson Learned: Understand and appreciate the risks surrounding you.

5. Keeping Focus at a South Africa Big Game Reserve

Without a doubt the most unique EH&S training program I have ever led was conducted at a big game reserve resort hotel for Colgate-Palmolive in South Africa. Extraordinary! CP brought in all their EH&S coordinators from their African operations, which are substantial. The training room had one wall, to my right as I recall, that was glass from floor to ceiling and looked out on the reserve. As we began the workshop, with the curtains open, all of us noticed gazelles, wildebeests, and zebras passing by. Sadly, I had to close the curtains and destroy the view. Not so much because the students were distracted but because the instructor was distracted.

One other interesting thing happened during the training week. One of the students told me that he had to take a video of his hotel room. His wife wanted to make sure that there were no strange women in his room tempting him. I'm not sure what a one-time snapshot video of a room tells you, but it meant a lot to his wife.

Lesson Learned: Two lessons here: (1) Enjoy the full experience, but focus on the task at hand, and (2) the wife is always right.

6. Searching for "The Little Mermaid" in Copenhagen, Denmark

The Little Mermaid statue is a Denmark national treasure located on a river bank in Copenhagen. On one trip we decided that this was a "must see" attraction that we needed to photograph. However, when we finally found it, there were several businessmen climbing all over it to

have their pictures taken in turn by their colleagues. This seemed a bit irreverent to us and did not make for a very good photograph. Doing a bit more research on the history of the statue, I discovered that it has been decapitated (and restored) at least twice. What a shame.

Lesson Learned: Respect international treasures.

7. Time Management in Mercedes, Argentina

We were at a chemical plant in Mercedes, Argentina, which is located about 100 km west of Buenos Aires. During the week-long visit, we noticed that no work could get done from before 12 noon to just after 2:00 P.M. as the employees and management went home for lunch. This frustrated the Americans who were used to either a 30-minute or working lunch. Yet, Friday was the biggest surprise. The site butchered an entire cow for their guests, and we had the longest and most carnivorous lunch I have ever experienced. They brought out each part of the barbequed cow, in order of its considered worth, from tongue, to organ meats, to the filet. Anthony Bourdain would have been in heaven; I was considering becoming a vegan.

Lesson Learned: If in Argentina, be ready to have a significant amount of downtime midday and be ready to eat beef.

8. Visiting a Client's Family Home in Japan

On a trip to Japan, our local client host invited us to his family home for dinner one evening. As we entered the modest but beautiful home with hardwood floors throughout, I noticed that there was a step up from the entry area to the living area. We were asked by the mother to remove our shoes before entering the living area, which seemed very civilized to me. So I placed my shoe on the step so that I could untie my shoelaces. I thought the poor mother was going to have a coronary on the spot. You see, even placing the sole of my shoe on the living area to remove it was considered blasphemy. Fortunately, after a sincere apology on my part, all was well. I'm sure I'll never forget the incident, and I suspect that the mother will never forget me.

Lesson Learned: Pay attention to the local customs.

9. Observing Religious Customs in Malaysia

I was leading an EH&S auditor training class at a manufacturing plant located in the outskirts of Kuala Lumpur, Malaysia. The class consisted of about 15 students, and we trained for two and a half days and conducted a practice site audit for two and a half days. During the audit, I noticed something special that I have only observed once in my career. Most of the class was Muslim, and five times a day the Muslim auditors would get out their prayer rugs, face Mecca, and pray. It put things in perspective for me; some things are more important than work.

Lesson Learned: Don't just tolerate religious rituals, appreciate and enjoy them.

10. Attending a Dinner Party in Johannesburg

When I visit plant sites thousands of miles away from home, I find that the locals are often very gracious towards their guests. One way they use to welcome visitors is to have a dinner party in their honor. This is indeed what happened on one trip to Johannesburg, South Africa. During the party, an Afrikaner woman mentioned that she had recently been on holiday in New York City. With all the amazing features of the City the one that most impressed her was that her hotel housekeeping staff actually consisted of men and women who were not black. This was a very different experience for her. And I should mention that she did not say this with any malice. To her it was just something unique and worthy of note.

Lesson Learned: There are times when you just simply have to bite your lip.

11. Being the Guest of Honor at Dinner in Shenzhen, China

Some years ago I was at yet another dinner party at a local restaurant in Shenzhen, China. As our group entered the restaurant, I noticed a corral at the left front, which contained live chickens, a live ostrich, and various other creatures. Down the main aisle of the restaurant were rows of aquariums containing live fish and other sea creatures. I learned very quickly that this stock of living things was our menu. And moreover, since I was their honored guest, I was to select who was to live and who was to be cooked and eaten, at least for that evening. Thankfully, the restaurant staff had not named any of the individual creatures Joe or Sally, which would truly have made me feel like an executioner. Oh, and by the way, I did give the ostrich and any animals with legs a reprieve.

Lesson Learned: As a guest, be ready for surprises.

12. Third-Shift Work in Notre Dame de Gravenchon, France

When doing community noise surveys at large petrochemical plants, it is important to distinguish the contribution of the plant from the community background noise. One way to do that is to do a fence line survey in the middle of the night when there is no traffic noise, in particular. That was what I was doing by myself one time in Notre Dame de Gravenchon, a French city with a population of ~10,000, 35 km inland from Le Havre. As I took my readings, I was interrupted and annoyed by the noise of police sirens. It turned out the gendarmes were actually looking for me. Someone in the community had called the police to complain that there was a strange guy walking around the neighborhood. My conversation with the officers was more sign language than anything else; I spoke only a very little French and they spoke no English. I managed to say "Esso" and then pulled out my noise meter and yelled the word "bruit," the word for noise in French, at it, and they watched the needle jump. I think they finally understood what I was doing. Nonetheless, I was escorted back to the rooming house where I was staying. My days of 2:00 A.M. noise surveys in rural France were over.

Lesson Learned: Don't be alone and where you shouldn't be at 2:00 A.M.

13. The Mexico Respirator

Some years ago, I was working at a chemical plant in the outskirts of Mexico City. We were auditing the plant from top to bottom. One task was to enter a major process building and

climb internal stairs to the roof. This seemed pretty straightforward until the site staff issued us half-face respirators as we stood at the building entrance. They told us that the respirators were required for building entry. So each of the visitors climbed the five flights of stairs donned with a respirator. I don't mind telling you that I was pretty winded when we reached the roof. It then occurred to me that I had just used personal protective equipment in an unsafe manner (e.g., no training, no fit testing, etc.). What was I thinking? I guess I wasn't.

Lesson Learned: Safety first, always!

14. The Singapore Culture

In my travels I have been to Singapore at least a half-dozen times. It is one of the prettiest and cleanest places I have ever been. For those who don't know, Singapore is a city-state located off the southern tip of the Malay Peninsula, about 90 miles north of the equator. English is one of four official languages and is the language used in business, unusual for Asia. I was told on one of my trips that Singapore is about the same size as Los Angeles but has more McDonalds. It is a very westernized environment, which can be very comforting to Americans. On one of my trips, there was some controversy over the "caning" of young men for what most would view as minor transgressions. For example, one of the local laws prohibited spitting on the sidewalk, and if one was caught, this was considered a canable offense. Now, I'm not normally a spitter, but as I walked along the main sidewalks in the business district one day all I could think to myself was "don't spit, don't spit, please don't spit!"

Lesson Learned: Be aware of and obey the local laws and customs.

15. Masks in South Africa

Today's economy is truly global. You can find expensive Italian shoes in the local mall; you don't have to travel to Rome. So what type of souvenir can you bring back that is truly unique to the country you visited and can't be found easily in the U.S.? I learned that there are at least two: Hard Rock Café t-shirts and tribal masks. When my two sons were growing up, they had enough HRC shirts to fill a closet; and indeed, because of that their mother finally put a halt to that practice. But collecting masks proved very fulfilling. I have about twenty masks on my office walls from pretty much everywhere. Each one is a special memory; one in particular. That one came from a street vendor in Johannesburg. I was there with a local colleague and decided on a particular mask that I really liked. My friend then began negotiations with the vendor in Swahili. Things got pretty heated, and I felt the need to step in. I asked my friend what the issue was and he told me that the vendor wanted the South African equivalent of a dollar for the mask and, on principle, he wasn't going above 50 cents. I quickly gave the vendor the dollar-equivalent and we went on our way.

Lesson Learned: Perspective is everything.

16. Lost Luggage at Charles de Gaulle Airport

I was working in Anzio, Italy, for a consumer goods company. Anzio is a coastal city about 35 miles south of Rome and was a city of great significance to the Allies during World War II. During the visit, the team was staying in Rome and commuting daily to Anzio. Having never been to Rome previously, I decided at the last minute to change my plans and stay an extra couple of days to visit the sites, and I'm glad I did, sort of. My new plan was to return home to Philadelphia via Paris' Charles de Gaulle Airport. All was going smoothly until I reached the departure gate at Charles de Gaulle. There was my plane, all set to go. And then the agent proceeded to tell me that my revised ticket had not been properly processed, and I could not board even though there were seats available. Sadly, I sat down and watched my ride home depart. I spent the next 12 hours in the airport and took a flight that evening not to Philadelphia but to New York's JFK airport. My luggage followed me home, 12 days later!

Lesson Learned: Be careful about making last minute travel adjustments when travelling internationally.

17. Returning from Australia

I was returning to the U.S. via San Francisco from a two-week long trip to Australia. The flight from Sydney to San Francisco is about 16 hours non-stop. That's two full work days sitting on a plane. I can assure you that there is nothing glamorous about that, even if you're sitting in business class as I was. As I recall, we landed early in the morning, and I was prepared for the typical customs re-entry hassle. But this time the hassle was like none other. As I handed the customs agent my passport he screams out at the top of his lungs—"YOU'RE THE ONE WE'VE BEEN LOOKING FOR!" If I weren't so exhausted from the trip, I probably would have run for my life. But thankfully I did not; I just stood there with a blank expression on my face. The agent then simply stamped my passport and told me to move along. To this day I'm not exactly sure what happened and why.

Lesson Learned: Keep your wits about you at all times. You never know.

18. Being Safe in South Korea

We were evaluating a chemical plant in South Korea. Our preparation was flawless, or so I thought. We had all the appropriate personal protective equipment to be safe in the plant: hard hats, safety glasses, hearing protection, safety shoes, and so on. However, there was one process area of the plant that required fire-retardant clothing, or Nomex. Not to worry, as the site had plenty of spare uniforms for visitors. There was one problem, however. There were no uniforms large enough for one person on the team, namely me. As a consequence, I was not allowed to enter that portion of the plant. It was embarrassing in so many ways.

Lesson Learned: Preparation down to the very last detail is paramount.

19. Ordering Train Tickets in Paris

I had three years of French in high school. Mind you, it was far from conversational French. It was more about learning vocabulary and syntax. So I figured that buying a train ticket by myself to Rouen at Paris' Gare du Nord, the busiest railway station in Europe, would be a challenge but not impossible. I was instructed by my French colleagues that I should go to the ticket window and say "première classe billet pour Rouen, s'il vous plaît?" or in English "first-class ticket for Rouen, please?"

The agent behind the window just shrugged his shoulders, implying he had no idea what I was requesting. I must have said the same phrase ten times with different pronunciations, with no success. There was now a long line building up behind me, with many becoming quite impatient, though none offered to help. Finally, I wrote down the statement on a piece of paper and gave it to the agent who now immediately understood what I needed. I was done in about a minute. Was it really my lousy pronunciation or was I being "punk'd"? Who knew?

Lesson Learned: Be ready to communicate your needs in more than one way.

20. Facing Off Against a Plant Manager in Thailand

Our team was conducting an evaluation of a chemical plant in Bangkok, one of my favorite cities in the world. Part of the review was to evaluate the performance of the on-site wastewater treatment plant. We discovered that the average daily biochemical oxygen demand (BOD) discharge concentration from the wastewater treatment plant was about 40 mg/l. The effluent standard listed in the permit was 30 mg/l. They were out of compliance and revising the permit was not an option.

The site manager told us that the plant is considered a model for the region and they have never been cited by the City of Bangkok. We also discovered that other companies up and down that stretch of the river, which is basically an open sewer, are discharging without treatment and BOD effluent concentrations were as much as several hundred mg/l. The local environmental agency had issued an environmental citizenship award to the site (a plaque was mounted in the lobby) and the agency director routinely brought other companies' management to the site to show them "how it should be done." It was going to cost about $250,000 to upgrade the plant to meet the 30 mg/l limit.

As auditors we had a dilemma. Do we report this as a finding to corporate management and recommend that the capital be spent to upgrade the facility? The plant manager was definitely opposed to spending the money on what he believed to already be a "best in class" operation for his city. It was not a comfortable situation, but we did record the issue in the report. Like most U.S. companies, this one had a corporate policy that they would operate in compliance with all environmental regulations wherever they operate in the world. Needless to say, the plant manager did not become my best friend.

Lesson Learned: Tell it like it is, regardless of the consequences. You will sleep better at night.

Overall Lessons Learned

With all my mishaps and near arrest in France, what I have learned that is most precious to me is that generally people at the grass-roots level around the world are pretty gracious towards their guests. I've felt welcome everywhere I've been. And that's a critically important lesson when all we seem to hear on the news is how badly we treat each other!

Note

1. Versions of this chapter were previously published by the author in the online *EHS Journal* as "Lessons Learned on International Assignments," December 27, 2012 and "Final Lessons Learned on International Assignments," April 28, 2013.

7

Classic Auditor Failures

How Do Auditors Mess Up?

"Only two things are infinite, the universe and human stupidity, and I'm not sure about the former."

—Albert Einstein

Introduction

Environmental health and safety (EH&S) auditing can be a challenging profession. It is fraught with ethical and technical dilemmas, the need to stay technically and functionally competent, and the need to have interpersonal skills and attributes that only Moses possessed. For example, International Organization for Standardization (ISO) auditing guidelines require auditors to be ethical, open-minded, diplomatic, observant, perceptive, versatile, tenacious, decisive, self-reliant, acting with fortitude, open to improvement, culturally sensitive, and collaborative.[1] Really, can any one person meet these expectations? Probably not.

As a consequence, auditors often stumble during the process, and based on numerous observations and experiences, this happens to both internal and third-party EH&S auditors. This chapter presents ten classic auditor failures and discusses ways to avoid them. The lessons are aimed principally at relatively new auditors, but the weary veteran should take note as well.[2]

1. Overall Approach—Defining Success

Wrong	*Right*
"We're there to find problems. It's almost like we're getting paid by the finding."	"We're there to help the site improve."

Generally, in conducting an independent audit of an operating facility, both internal and third-party auditors are asked to be objective in their evaluation, are expected to work with the site in a supportive way to achieve improvement, and occasionally are asked to help the site prepare for a regulatory agency inspection. Auditors do not always appreciate the nature of their role and how their role differs from that of an agency inspector. There have been cases in which auditors have stated in the opening conference, "We get paid by the finding." Although this is said in jest, site staff clearly will not appreciate the humor. Other auditors visibly relish uncovering deficiencies and have informal awards among themselves for the "finding of the audit." Still other auditors are sensitive about being perceived as not having exerted sufficient effort if a finding threshold is not reached and about hearing "you just did not look hard enough."

There should be no joy in identifying shortcomings. The role of auditors is to be professional, independent, objective, and without bias and to let the facts dictate the outcome. An audit with very few findings does not necessarily mean the auditors didn't do their jobs. It more likely means that the site is well run.

2. Preparation and Planning

Wrong	*Right*
"We'll hit the ground running on Monday; real-time planning is the way to go."	"We must be sensitive to the site's scheduling needs; we should set an agenda prior to the audit and be on time for all events."

Audits involve numerous interviews, daily meetings, opening and closing conferences, and other activities where on-time arrival is critical. In some cases, auditors have arrived late to a site because of air travel problems, have been late to interviews because they didn't know where the interview was to take place, have come late for daily meetings because they underestimated the time an interview would take, and so forth. The auditor's job is to be consistently on time for these activities, even if site staff is not. In other cases, when agendas are not prepared in advance, key site personnel are not available during the audit because of conferences, vacations, or other scheduled events. Not having access to key personnel can greatly reduce the value of the audit. Finally, if the audit team does not work through the agenda in advance of the audit, they could miss key observation opportunities, such as tanker truck loading or unloading events, or they might not arrive with appropriate personal protective equipment, which could prevent them from visiting key areas of the site.

At all times, auditors should know where they have to be and when. Auditors should understand the schedules of key site staff and work through potential conflicts in advance of the audit. Auditors should make necessary preparations to arrive on site ready and able to visit key areas and interview key personnel. For opening conferences in particular, auditors should know how long it will take to get to the site from the hotel, what security or safety orientation will be required when they get there, and how long it takes to get to the meeting room from the "front gate." Being late for the opening conference does not set a good precedent. Auditors should know how to get in touch with the site EH&S coordinator, so that if they are running late, the site staff knows about it. Site staff shouldn't have to come looking for the auditors.

3. Distractions

Wrong	*Right*
"I have a lot of other work to do while I'm on the audit; I hope we have a good cell connection. I also need to be on the Internet with my laptop."	"This is going to be a very long week. I have some other work that I will have to do in the hotel at night. I hope the hotel has a fast wireless Internet connection."

These days, auditors are minimally outfitted with a laptop and a smartphone. In a world where every e-mail and communication is "urgent," auditors often feel a strong urge on an audit to use (or overuse) these devices to "stay in touch." In some cases, auditors have taken a call during the opening conference; in other cases, auditors have spent more time on the phone or with e-mail during the audit day than on the audit itself. How do we know that this very inappropriate behavior has happened? Because we have observed it first hand and have received complaints from clients about auditors not being dedicated to the audit while on-site.

All auditors must be discrete in the use of communication devices while on-site during an audit. Checking messages and conducting other work should be done only during lunch breaks (with permission) or back at the hotel. Auditors should be fully focused on the audit during the time spent on-site. They need to turn their phones off or to mute status.

4. Auditor Posture Toward the Site

Wrong	*Right*
"They will hide things from us; we must be suspicious."	"They will be open and candid; still, we should be thorough."

"Trust but verify" was a signature phrase adopted and made famous by President Ronald Reagan. Reagan frequently used it when discussing U.S. relations with the Soviet Union. This phrase translates very well to EH&S audits. It is quite rare that a site will purposely hide issues from an audit team. On the other hand, if an auditor does not ask about an issue that the site might be facing, there are certainly cases where the site staff will not volunteer the information. Auditors must be rigorous in the evaluation but not necessarily suspicious of the site staff's behavior.

5. Findings Ownership

Wrong	*Right*
"The findings are mine; the site doesn't have to agree."	"We can negotiate the findings but the final call is mine."

There is a saying that "all things are negotiable." It's not clear that it applies to death and taxes, but it should apply to audit findings. For auditors, it makes perfect sense to listen to the site's point of view on the technical merit of any particular finding and to do so sincerely. It should also apply when auditors are required to assign risk levels to findings (e.g., severe, moderate, or minor) or to classify a finding as a repeat. In auditing, most issues are far from black and white. Auditors should keep in mind that the word *audit* comes from the Latin word *audire*, which means *"to listen."*

6. Findings Credibility

Wrong	*Right*
"Findings can be based on my opinion; after all, that's why they're paying me."	"My findings should be defensible; they should be based on specific requirements."

The U.S. Environmental Protection Agency (EPA) describes compliance auditing as "a systematic, documented, periodic, and objective review by a regulated entity of facility operations and practices related to meeting environmental *requirements*."[3] These requirements can be regulatory in nature or self-imposed performance standards. Auditors should be principally evaluating the site's status against these requirements and should avoid drawing broad conclusions based on personal opinions about what is appropriate. Notwithstanding this comment, auditors are sometimes asked to provide recommendations on how to improve performance generally, regardless of applicable requirements. This request is perfectly acceptable but these recommendations should be clearly distinguished from findings of deficiency with respect to compliance, and the focus of the audit should remain on compliance with requirements.

Auditors should understand precisely, and in advance, the expectations with respect to findings of deficiency and opportunities for improvement. If both are to be addressed, auditors should make sure that the basis for the observation, in either case, is fundamentally sound and defensible. Auditors will be challenged.

7. Day-to-Day Communications

Wrong	*Right*
"We are secretive with the findings until the closing conference. We want to see the expression on their faces when we discuss the deficiencies."	"We communicate openly and freely throughout the on-site process. We want to get it right."

The EH&S on-site process should be transparent throughout. It benefits all concerned when potential findings are communicated as soon as they are observed. In numerous cases, an auditor believed that he had a "stone-cold" finding only to discover in the closing conference that he had misunderstood a site person's comments or had not interviewed the right person. This situation can be quite embarrassing to the audit team. Sometimes, the fear of sounding too confrontational keeps auditors from raising issues early on in the process, but there are ways to open up the discussion of a potential finding without sounding too confrontational. One auditor, when initiating such a discussion with site staff, uses the phrase, "Would it be fair to say. . . ."

8. Scheduling the Closing Conference

Wrong	*Right*
"The closing conference will be scheduled based on my travel wishes."	"The closing conference will be scheduled to ensure that we're done and the plant manager can attend."

A closing conference can make or break an EH&S audit. It is the culmination of the on-site work and typically involves the site manager and key site staff. Key corporate staff and legal counsel might also be in attendance, either in person, by phone, or by webinar. The logistics of scheduling the conference can be quite a challenge, particularly when remote participants are in different time zones. Auditors should be flexible and accommodating and schedule the conference to ensure maximum participation. If that means the audit team members miss their previously scheduled flights, then so be it; that's why they invented Saturdays. Auditors who constantly look at their watches during the closing conference could be considered rude by some. Audits are not typically "a walk in the park."

9. Communicating Post–On-site Findings

Wrong	*Right*
"It's okay if there are new findings in the report that were not discussed while the team was on-site."	"New findings will be very unlikely; if this situation occurs, we will alert the site before we issue the draft report."

It is the goal of every audit team to be done with the audit proper before leaving the site. However, this does not always occur; further interpretation of a regulation might be needed, data or records might not have been readily available at the site, and so forth. Should post–on-site evaluations result in one or more new findings, it is only courteous to let the site know about these developments before the draft report is issued. Auditors should follow a "no surprises" philosophy.

10. Follow-on Work (for third-party auditors only)

Wrong	*Right*
"This audit is going to be great! I'll bet there's $100,000 in follow-on work for us."	"We need to be clear about how we handle any potential follow-on work. Let's wait until the audit is over and see what the client's expectations are."

Audits can be a source of follow-up work for third parties. After all, the objective on an audit is to identify all the EH&S deficiencies at the site. However, if auditors appear too eager, which they sometimes do, to help the site fix all the identified deficiencies, it can come across as "ambulance chasing." To be sure, clients have complained about this situation to third-party consultants, particularly when auditors are already proposing scopes of work on the first day of the audit.

Third-party auditors need to understand the rules with regard to follow-on work on each audit assignment. Some clients are more than happy to have the third party help the site correct the deficiencies. Other clients actually prohibit third parties from being involved in the fixes because they believe that this involvement compromises the independence and objectivity of the audit. For each client, the third party should know the rules before the on-site visit. In situations where the auditor is allowed to help the site with the fixes, the auditor should show restraint while on-site and should wait until after the closing conference before discussing follow-on opportunities.

Conclusion

Wrong	*Right*
OVERALL . . . "We were relentless in identifying, justifying, and reporting all of the site's deficiencies."	OVERALL . . . "We made a substantial difference in improving the safety and environmental performance at the site."

The overall value of an audit should not be judged by how many findings appear in the audit report. Rather, the focus should be on the value added to the site and the reduction in risk realized as a result of the audit. The value of the audit is maximized when auditors collaboratively work with sites to reduce risks.

Notes

1. ISO 19011:2011, Guidelines for Auditing Management Systems, Section 7.2.2.

2. A version of this chapter was previously published by L.B. Cahill and Robert J. Costello, PE, CPEA, Esq. in the on-line *EHS Journal*: "Classic Auditor Failures," June 30, 2012.

3. "Environmental Auditing Policy Statement-Introduction," U.S. Environmental Protection Agency, *Federal Register*, Vol. 51, No. 131, July 9, 1986, p. 25006.

8

Repeat versus Recurring Findings on EH&S Audits

Should Management Treat Them Differently?

"Insanity is doing the same thing over and over again and expecting different results."

—Albert Einstein

Repeat	**Recur**
Transitive verb	Intransitive verb
To go through or experience again (had to *repeat* third grade)	To occur again after an interval: occur time after time (the cancer *recurred*)
Source: Merriam-Webster's Dictionary	

Introduction

Many organizations have been conducting environmental health and safety (EH&S) audits for decades now, and as auditors revisit sites on a 2- or 3-year cycle, one of the most frustrating aspects of the audit outcomes is that many of the same or similar findings of non-compliance seem to arise time and time again. This situation is obviously troubling because it implies that site managers are not addressing the findings in a way that (1) results in permanent fixes and (2) truly changes the culture of the organization.

Hence, in order to minimize or even eliminate the occurrence of repeat findings, corporate management often puts in place a program of punitive measures or sanctions that are directed at site managers. This approach can be problematic for any number of reasons:

- It can change the tone of an audit program from independent and supportive to confrontational and combative. Site managers are never comfortable with repeat findings when a penalty might be incurred. The only possible exception is when the problem has persisted because of a lack of corporate or business unit funding for the necessary capital improvements.
- Perhaps more importantly, not everyone has the same idea about what is or is not a repeat finding. No consensus definition has been developed to help classify a finding as a repeat, which can cause serious disputes between site managers and the audit team. Moreover,

where definitions have been developed, they are often not considered satisfactory to all concerned stakeholders. For example, in some audit programs, a repeat finding classification is given solely if the identical regulatory citation applies, regardless of whether the actual deficiency is precisely the same. In other audit programs, the technical deficiency must be almost identical to the one identified on the previous audit. In still other programs, classifying a finding as a repeat is basically at the auditor's discretion.

This brief chapter focuses on the issue of defining repeat findings.[1] A "repeat finding" classification must be defined for an EH&S audit program so that everyone is working from the same rule book. This will increase understanding across the board and hopefully minimize disputes.

Repeat findings are typically considered to be serious and justifiably receive significant management attention. However, before labeling something a "repeat," auditors must focus on what actually caused the finding to occur. The question really is: "Did a breakdown in a management or control system cause this failure, or was it simply an *isolated* case of a similar nature?" In other words—is this a repeat finding or a recurring finding?

What Is a Repeat Finding?

A repeat finding can be defined as one of the following:

- A finding that was identified in the previous independent audit for which a corrective action has not been completed as planned
- A finding that is substantially similar in nature to one that was identified in the previous independent audit

An example of an environmental audit finding that would clearly be considered a repeat would be the case of a site that on an initial audit is determined to be operating without a required air or wastewater permit and on the follow-up or subsequent audit still does not have the required permit. One factor that might color the repeat classification would be if the site had applied for a permit promptly but the issuing agency had not yet responded. The auditor would then have to decide whether the site follow-up efforts with the agency, over a three-year period, were earnest and substantial (e.g., face-to-face meetings). If they were not, then there is a strong implication that the corrective actions were insufficient to correct the deficiency, and this could legitimately be called a repeat finding.

An example of a repeat safety finding would be a case such as the following: On the initial audit, a guard that protected workers from coming into contact with a high-speed belt was observed to have been removed from a major piece of rotating equipment. Operators claimed that the guard "got in the way." After the audit, the site reported in its corrective action plan that the guard had been reinstalled. Three years later, on the next audit, the guard was once again found to have been removed. This is another repeat finding.

What Is a Recurring Finding?

A finding that would probably not be considered a repeat finding is seen in the following example of an environmental audit of a wastewater treatment plant. Assume that 10 parameters (e.g., pH, BOD, and TSS) with daily effluent limits are listed in the discharge permit and that the audit review period is three years. This means that roughly 10,000 data points must be evaluated. Now assume that, on an initial audit, a small number of minor excursions of pH or BOD occurred over the three-year audit period. Next, assume that on the subsequent audit three years later, it was determined that a small number of minor excursions of pH or BOD had occurred as well. Should the second set of excursions constitute a repeat finding if the management of the wastewater plant is otherwise found to be fundamentally sound? This is probably not a repeat finding, since wastewater treatment plants rarely operate in compliance with the permit limits 100 percent of the time.

A second example of a finding that might not be considered a repeat could be found on a fire safety audit at a large manufacturing site. Auditors, if they look long and hard enough, can almost always find an issue with inspections and maintenance of portable fire extinguishers. Should a missing tag on one fire extinguisher out of a universe of several hundred constitute a repeat finding if another extinguisher was without a tag on the previous audit? Again, if the management system is otherwise sound, the answer is "probably not."

Similar situations can occur in other compliance areas where the universe of events or items to be audited is also quite large (e.g., material safety data sheets, hazardous waste manifests) and the likelihood of isolated failures is high. Auditors should think of these as recurring findings instead of repeat findings.

Conclusion

Identifying true repeat findings is a critical component of any EH&S audit program, especially given that repeats can be interpreted as "willful and knowing" violations by regulatory agencies. However, there is a difference between "repeat" and "recurring" findings: repeat findings result from the breakdown of a management system or control, whereas recurring findings are mostly isolated occurrences that can happen in the best of programs. It is important for auditors to understand the difference between the two and ensure that sites are not punished or sanctioned by the repeat classification when the controlling system is otherwise fully implemented and effective.

Note

1. A version of this chapter was previously published by L.B. Cahill and Robert J. Costello, PE, CPEA, Esq. in the on-line *EHS Journal*: "Repeat vs. Recurring Findings on EHS Audits," March 31, 2012.

9

Environmental Audits versus Health and Safety Audits

What's the Difference for Auditors?

"Try to learn something about everything and everything about something."

—Thomas Huxley

Introduction

The great majority of environmental health and safety (EH&S) professionals come from a background that is either primarily environmental in focus or primarily health and safety based. As the EH&S auditing profession has evolved, companies have integrated environmental topics with health and safety topics (and sometimes with others such as management systems or sustainability) on site audits, and many auditors must work outside their traditional comfort zones, whether in terms of education or experience. For example, at smaller sites, companies often expect one auditor to be able to cover all EH&S topics. Even at larger sites, where there is a team of auditors, the expectation is that each auditor should be able to contribute on topics that might be more involved than anticipated. This situation holds true for both internal and external auditors.

Aside from the obvious variations in the content of regulations and requirements, environmental audits and health and safety audits differ to varying degrees in other ways. These differences must be recognized and understood in order for auditors to conduct audits effectively and efficiently. This chapter discusses 12 of these differences, with the goal of helping auditors, especially new auditors, understand the subtle and not-so-subtle differences between the two types of audits and making them comfortable working on topics for which they are not subject-matter experts.[1]

Volume of Federal Regulations

First and foremost, the United States currently has about nine times as many pages of federal environmental regulations as it does pages of health and safety regulations (see Chapter 2). However, many environmental regulations will not apply to a particular audit. For example,

64 percent of the environmental regulations are related to air emissions, and many of these apply only to particular sources. This disparity in volume does not necessarily mean auditors can more easily understand health and safety requirements than environmental requirements. It does, however, mean that the body of knowledge is much larger for environmental audits, and environmental auditors must understand which regulations apply to a specific audit and which do not.

Volume of State Regulations

A striking difference in environmental regulations versus health and safety regulations is also found at the state level, where regulatory agencies must adopt federal requirements as a regulatory minimum and can, at their discretion, promulgate only state requirements that are more stringent than the federal requirements. Most states were delegated the authority to implement federal environmental programs and often have developed even more stringent environmental requirements. Some states have developed environmental programs in parallel with federal programs, which requires auditors to audit against both the federal and the state requirements for a specific program (e.g., New Jersey developed discharge prevention regulations parallel to the federal spill prevention rules). A common environmental auditor pitfall is failing to audit against state-specific environmental requirements. Conversely, except for California and a few other states, more stringent state health and safety regulations are rare, so the focus is more typically on the federal requirements.

One way to evaluate the relative differences in regulation proliferation at the state level is to review state audit protocols offered by commercial providers. For instance, Specialty Technical Publishers (STP) offers a complete set of federal and state EH&S audit protocols that include state regulatory difference summaries for all 50 states.[2] These summaries include only those requirements that are particular to the given state. Interestingly, the average number of pages for the environmental difference summaries is 66, whereas that for the health and safety summaries is only 17, despite California's high number (see Figure 9.1). This reinforces the assumption that state-initiated environmental regulations are much more common and voluminous than their health and safety counterparts.

FIGURE 9.1
Top Five States—STP Regulatory Summaries

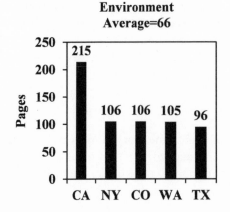

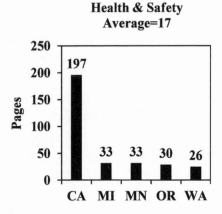

Completeness of Regulations

Differences between environmental regulations and health and safety regulations also exist with respect to regulation completeness. On the environmental side of the ledger, the regulations are more likely to be self-contained, although on occasion auditors need to review the preamble to the regulation, published questions and answers, guidance documents, and so forth to better interpret the meaning of the requirement. On the health and safety side, many letters of interpretation help to define the requirements. As a measure of how many letters there are, a search for "letters of interpretation" on the U.S. Occupational Safety and Health Administration (OSHA) website resulted in 55,000 hits. If OSHA issues a letter of interpretation, the regulated entity is expected to know its contents, especially if there is some ambiguity in the regulation. Companies have been cited because they did not know the rigor of the expectation defined in letters of interpretation, especially in the Control of Hazardous Energy standard. Thus, identifying individual letters of interpretation is not a straightforward exercise but is nonetheless necessary.

Regulatory Approach

Many environmental regulations include descriptions of not only what to do but how to do it. For example, the U.S. Environmental Protection Agency (EPA) requires not only that containers of hazardous waste be labeled but also that they meet the following requirements:

1. The initial date of waste accumulation must be clearly marked and visible on each container.
2. Each container or tank must be labeled or marked clearly with the words "Hazardous Waste."
3. No waste may be stored for more than 90 days unless an extension has been granted by the EPA.
4. Containers must be closed except when wastes are added or removed.

Conversely, health and safety regulations often include only the "what to do." For example, OSHA requires that a written confined space permit program must, among many other requirements, do the following:

1. Specify acceptable entry conditions appropriate for the types of permit spaces present at the facility.
2. Develop procedures for isolating the space and, as appropriate, purging, inerting, flushing, or ventilating the permit space to eliminate or control atmospheric hazards.
3. Develop effective methods to restrict pedestrian and vehicular access to the entry areas to protect entrants from external hazards.

For an auditor, the task of verifying compliance would typically be much more straightforward with the more prescriptive EPA regulations.

Number of Audit Topics

Although there are nine times as many pages of federal environmental regulations as there are health and safety regulations, a health and safety audit typically covers twice as many audit topics (see Table 9.1). As a result, the health and safety auditor often needs to cover more topics than the environmental auditor on a given audit.

The topics shown in italics in Table 9.1 tend to be easier for auditors who are not subject-matter experts to audit. Often, these topics are easier because they emphasize written programs, training, and inspection records, rather than detailed knowledge of the intent of the regulation or knowledge of whether an answer to an audit question is credible. These topics are good ones to start with for auditors who want to broaden their capabilities and cross over to the other side—but they are always to be approached with caution.

Overall Focus and Principal Means of Verification

The three core evidence-gathering activities on EH&S audits are records review, interviews, and physical inspections. On environmental audits, written records are critically important and, as a result, receive considerable attention from environmental auditors. These records can include regulatory permits and plans (e.g., wastewater permits, air permits, spill response and contingency plans), written procedures, inspections, data reports, and the like. Conversely, on health and safety audits, although the review of records (e.g., permits, training, written programs, and inspection logs) is certainly important, observation of workforce behaviors and interviews, in particular, are relatively more important in verifying health and safety compliance.

TABLE 9.1
Typical Topics on EH&S Audits

Health and Safety Topics (24)	Environmental Topics (12)
Blood-borne pathogens	Air emissions
Confined space entry	Community Right-to-Know
Control of hazardous energy	Drinking water
Electrical safety	Hazardous wastes
Emergency response	*Medical wastes*
Ergonomics	*Pesticides*
Fall protection	*Solid wastes*
Fire protection	Spill containment
Flammable storage	*Storm water*
Hazard communication	Toxic substances
Hearing conservation	Wastewater
Hot work	*Universal wastes*
Industrial hygiene	
Lifting devices, cranes, and hoists	
Machine guarding	
Means of egress	
Medical and first aid	
Personal protective equipment	
Portable hand tools	
Powered industrial trucks	
Record keeping	
Respiratory protection	
Ventilation	
Walking and working surfaces	

Note: The topics in italics are more suitable for crossover auditors

Availability of Records

All environmental records are typically available to auditors; however, not all health and safety records are. Certain personnel records are considered confidential (e.g., audiograms conducted as part of a hearing conservation program, pulmonary function tests conducted as part of a respiratory protection program), and they cannot be released to auditors without employees' written approval. This situation can make conducting the occupational health portion of a health and safety audit challenging. It may be necessary to have a plant nurse or other professional review the records and answer the auditor's questions while reviewing the records instead of allowing the auditor direct access to the records.

Approach to Permits

An intriguing difference in the approach to permits exists between environmental and health and safety topics. There are, of course, numerous agency-issued environmental permits, and reviewing these permits is vital to the verification process. In contrast, there are no agency-issued health and safety permits, but there are health and safety permits nonetheless. These permits are internal to the organization and typically must be issued when a particular activity takes place (e.g., confined space entry, hot work). Evaluating whether these permits contain all the appropriate elements and have been completed correctly is an important part of the verification process.

Governing Programs and Procedures

Records review on EH&S audits can be an intimidating exercise. However, certain key documents can provide a road map for verification. On the environmental side, these documents include both agency-issued permits and written plans (e.g., storm water pollution prevention plan, spill plan, hazardous waste contingency plan) that have been prepared by the facility. In many cases, the key strategic documents for health and safety include actual written programs that define the particular approach for meeting the requirements of the standard. For example, confined space entry, hazard communication, respiratory protection, and control of hazardous energy require a written program that governs the activity. These written programs, when required and available, are perfect documents for quickly understanding a program and for developing a verification strategy.

Agency Regulatory Databases

It has become standard audit practice for auditors to access U.S. government databases before the site visit in order to gather regulatory and compliance information about the site. Presently, three databases provide this type of information; two are managed by the EPA, and one is managed by OSHA. They are as follows:

- **EPA Envirofacts.** Envirofacts is a single point of access to selected EPA environmental data. This website provides access to several EPA databases that contain information about environmental facilities that may affect air, water, and land anywhere in the United States (www.epa.gov/enviro).
- **EPA ECHO.** ECHO provides searches of EPA and state data for more than 800,000 regulated facilities. ECHO integrates inspection, violation, and enforcement information for

the Clean Water Act, the Clean Air Act, and hazardous waste laws. ECHO also includes data on the Safe Drinking Water Act, the Toxics Release Inventory, the National Emissions Inventory, and water quality (www.epa-echo.gov).

- *OSHA IMIS.* The "establishment search" function enables auditors to search the OSHA Integrated Management Information System (IMIS) enforcement database by establishment name. This database contains information on more than 3 million inspections conducted since 1972. The database is updated daily from more than 120 OSHA area and state 18b plan offices (www.osha.gov/oshstats).

Auditors need to become familiar with these databases and use them as a resource on every audit conducted in the United States. Otherwise, the audit will not meet "generally accepted practice" expectations currently in play.

Classification of Audit Findings

Many audit programs require auditors to classify each finding by the level of risk it poses to the organization. Some schemes use the terms significant, major, and minor. Others classify findings as high, medium, or low risk, and still others designate findings as level I, II, or III. Definitions are provided for each risk category, but all schemes require judgment on the part of the auditor. For environmental findings, these classification systems work relatively well. For safety findings, however, they can sometimes prove problematic. For certain issues (e.g., a machine guard missing from a high-speed belt in a remote location), auditors are more likely to envision a maximum worst-case scenario that would result in an injury or fatality, even though the likelihood of the event is extremely low. This can make assigning a risk level more challenging.

Financial Consequences of Noncompliance

It seems rather odd that, historically, the financial consequences for environmental incidents have dwarfed those for health and safety incidents, although some people would say this is changing. For example, the BP Texas City refinery explosion in 2005, which killed 15 contractors, resulted in the largest fines ever assessed by OSHA, a total of more than $100 million in two separate enforcement actions (2005 and 2009). However, the financial consequences for BP of the 2010 Deepwater Horizon release, as opposed to the explosion, have been in the tens of billions of dollars.

One reason for this disparity in financial consequences might be found in the relative budgets of the EPA and OSHA. The EPA's 2015 budget is 14 times that of OSHA (see Chapter 2). In fact, the EPA's enforcement budget alone is 40% larger than OSHA's entire budget!

Conclusion

In most ways, conducting environmental audits is quite similar to conducting health and safety audits. However, the differences between the two types of audits should be understood and incorporated into an audit strategy. For environmental auditors who are being asked to audit health and safety topics, the most important differences are the following:

- Interviews and observation of behaviors are key to verifying compliance.
- Twice as many topics must be covered in the same period of time.
- Becoming adept at identifying key "letters of interpretation" is important.

Conversely, health and safety auditors who are asked to audit environmental topics should note the following important differences:

- Understanding what regulations apply must be done as part of the scope definition.
- Assessing state requirements is key to evaluating compliance.
- A detailed review of records is necessary.

As the EH&S audit practice continues to mature, auditors will be expected to be more facile across a broader range of topics and will need to recognize both the similarities in and differences between environmental audits and health and safety audits. However, auditors should always be cautious about attempting to audit potentially high risk areas without having the knowledge and experience necessary to conduct a thorough and insightful audit.

Notes

1. A version of this chapter was previously published by L.B. Cahill and Robert J. Costello, PE, CPEA, Esq. in the on-line *EHS Journal*: "Environmental Audits versus Health and Safety Audits," March 6, 2012.

2. Environmental and Health and Safety State Differences Summaries and Checklists, Specialty Technical Publishers, Vancouver, BC (www.stpub.com).

10

Using Risk Factors to Determine EH&S Audit Frequency

How to Best Manage Audit Resources

"Time is the scarcest resource, and unless it is managed nothing else can be managed."

—Peter Drucker

Introduction

Establishing appropriate EH&S audit frequencies for sites and facilities considered to be part of the auditable universe can be a trying exercise. Auditing high risk operations too infrequently can lead to unwanted surprises due to a lack of oversight and governance. On the other hand, auditing too frequently can be costly, lead to a feeling of unbearable oversight by the audited community, and can eventually compromise the effectiveness of the program, to say nothing of the reception given to audit teams when they arrive at a site.

On one environmental audit some years ago at a large chemical plant in California, the plant manager's first words in the opening meeting were: "I can't believe we're being audited yet again. Do you know that we've been audited or inspected over 75 times this quarter alone?" The team leader (and the company) failed to recognize that this site, a major military contractor for the U.S. government, was receiving regular "attention" from the corporate audit group, customers, and regulators on topics such as finance, security, environment, health and safety, process safety, transportation, and so forth.

So what's the right frequency and where can one go for guidance? This chapter attempts to help with answering those questions.[1] Discussions include:

- Expectations of regulatory agencies and professional organizations
- An overview of how to assign risk factors to auditable sites to establish frequency
- Examples of how nine companies address audit frequency

Although there is no perfect solution that applies to all cases, the general approach and examples from nine companies can be used to tailor a solution.

Expectations of Regulatory Agencies and Professional Organizations

No agency or professional organization prescribes exact frequencies for EH&S audits. Those that do address the subject say that audit frequency should be based on risk (Auditing Roundtable, Board of Environmental, Health and Safety Auditor Certifications [BEAC], The Institute of Internal Auditors [IIA]) and others say the frequency should be "periodic" (ANSI/AIHA Z10, ISO 14001, OHSAS 18001, U.S. Environmental Protection Agency, U.S. Sentencing Commission) (see Table 10.1). Although it should be noted that most of the external guidelines are notably silent on specific quantitative expectations for audit frequency, the commonly held expectation for audits is that major facilities generally are audited no less frequently than once every two to three years.

Assigning Risk Factors to Auditable Sites to Establish Frequency—An Overview

There are, of course, many approaches to ranking facilities by risk and other factors and subsequently setting frequencies based on those rankings. Generally, site risk factors can be viewed as two-fold: inherent and external. First, there are indeed inherent risks of operation, which can

TABLE 10.1
Audit Frequency Expectations of Regulatory Agencies and Professional Organizations

Organization	Document	Frequency Expectation
ANSI/AIHA	Occupational Health and Safety Management Systems Standard (ANSI/AIHA Z10-2005)	"periodic"
ASTM International	Standard Practice for Environmental Regulatory Compliance Audits (E2107-06)	Silent on the subject.
Auditing Roundtable	Standard for the Design and Implementation of an EH&S Audit Program (1996)	Frequency "based on existing or potential EHS impacts, taking into account such factors as level of EHS risk. . . "
BEAC	Performance and Program Standards for the Professional Practice of EH&S Auditing (2008)	". . . facilities which pose the greatest risk to the company are audited earlier in the cycle, or at more frequent intervals, than other facilities which pose less risk."
IIA	Standards for the Professional Practice of Internal Auditing (1997)	Primarily risk-based
ISO 14001	Environmental Management Systems (ISO 14001:2004(E))	"periodic"
ISO 19011	Guidelines for quality and/or environmental management systems auditing (ISO 19011:2002(E))	Silent on the subject.
Occupational Health and Safety Assessment Series (OHSAS)	Occupational Health and Safety Management Systems Standard (OHSAS 18001:2007)	"periodic"
U.S. EPA	Environmental Auditing Policy Statement (1986)	"periodic"
U.S. Sentencing Commission	Sentencing Guidelines Manual, Effective Compliance and Ethics Program (§8B2.1) (2010)	"periodic"

involve the materials handled, the age of the facility, and the complexity of the process. These risks are important but perhaps more controllable than the second class of external risks that may include the company's compliance history, the community and environmental setting, and the state/local agency's regulatory stringency.

If one views these two classes of risk in concert, as in Figure 10.1, a facility-by-facility risk evaluation can be conducted. We can find fairly large facilities, such as Facility A, that pose high risks, and efforts can be undertaken to reduce both inherent and external risks to move this facility into either a relatively safe or controllable situation. Such efforts might include increasing measures to reduce noncompliance (i.e., increased audit frequency) or investigating the possibility of materials substitution. For another facility, such as Facility B, which poses only modest inherent risk, but is in so unstable an external environment that it is vulnerable to unwanted surprises, a public relations or compliance improvement program can be developed that will move the facility to the "relatively safe" category.

In much the same way, all facilities or units can be evaluated for their relative risk potential. Corrective action programs can then be fine tuned to address the nature and extent of the risk with the most cost-effective solution. One of these solutions could be the development of an EH&S audit program that uses the material and facility risk assessment techniques discussed previously as priority setting tools. As shown in Table 10.2, the scope, frequency, and resource commitment can be assigned based on an estimate of risks posed by the facility and the materials handled at the facility. In this way, resources are committed cost effectively. That is, the number of auditors and the frequency of audits are aligned with the relative risks posed by each site.

Table 10.2 can also be used as a resource planning tool. Once the company's inventory of facilities to be audited is established and a frequency, audit duration, and team size is assigned

FIGURE 10.1
Assessing Risk in a Multi-Plant Environment

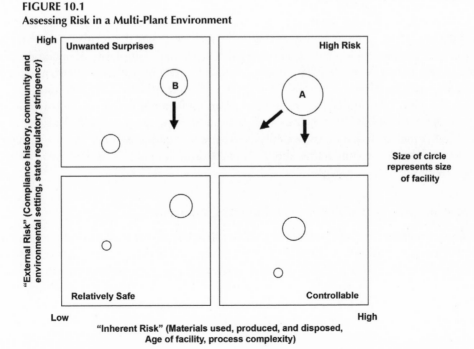

TABLE 10.2
Establishing Audit Frequencies

Site	Risk Class*	Frequency	Duration	No. Auditors
Large Ag Chem Processing Plant	Very High	Annual	One Week	5
Machine Tool Plant	High	Every 2 Years	3.5 Days	3
Drum Storage Facility	Medium	Every 3 Years	1.5 Days	2
Truck Terminal	Low	Every 4 Years	1 Day	1

*Based on incident history, materials handled, process complexity, environmental setting

to each facility, the manpower loading for field audits can be determined for any given year. Further, if the number of field hours is increased by 50 percent or so to account for audit preparation and report writing, the result should indicate full labor cost accounting for the program, except for management and administration time. Compiling this information on a spreadsheet will allow the program manager to manipulate critical factors, such as audit frequency, to determine the financial or budgetary impacts of increasing or decreasing the frequency.

Examples of How Nine Companies Have Addressed Audit Frequency

Provided below are specific examples of how nine companies establish audit frequencies. The information is taken from actual corporate audit procedures. It is clear that "no one size fits all." The idea is to present some options that allow an organization to design a tailored program, drawing from the most applicable attributes of each example.

1. A Large Pharmaceutical Company

No facility within the audit "pool" is to be evaluated at a frequency of longer than four years. Schedules are set by using a "criticality matrix," which evaluates the relative risk of facilities using criteria such as employee population, regulatory climate, complexity of operations, facility location, accident rates, site EH&S resources, extent of facility self-assessments and the like. Nominally, audit frequency is established using the schedule presented in Table 10.3.

Each site within the auditable pool is assigned an initial audit frequency. This frequency is adjusted annually and at the conclusion of each audit, depending on a re-assessment using the criticality matrix. Based on the established frequency and the number of sites in the pool, the program should be conducting about 27 evaluations annually. This analysis is shown in Table 10.4 on the following page.

2. A Medium-sized Mining and Minerals Processing Company

The Audit Program Director will develop the audit schedule for each year and assign audit team leaders from his staff of managers. The frequency at which a site is audited, how long the audit will take, and how many auditors will participate will be based on the perceived risks of the site. An evaluation of these parameters will be made by the end of each year by the Audit Program Director in consultation with Group EH&S Coordinators based on the criteria presented in Table 10.5. The criteria will be used as a guide not as a quantitative

TABLE 10.3
Audit Frequency Schedule Based on Risk

Relative EH&S Risk	Audit Frequency
High	Every 18–24 months
Medium-High	Every 26–32 months
Medium-Low	Every 34–40 months
Low	Every 42–48 months

TABLE 10.4
Number of EH&S Evaluations Required Each Year Using the Criticality Matrix

	Criticality Risk Factors				
	Low	Medium Low	Medium High	High	
Factors	<1.5	1.5–1.9	2.0–2.5	>2.5	Totals
No. of Facilities with Factor	11	31	35	3	80
Max. Allowed Frequency (mos.)	48	40	32	24	–
Required Facilities/Year	2.8	9.3	13.1	1.5	~26.7

TABLE 10.5
Risk Factors Used in Assigning Site Audit Frequency

Site Characteristics	Category I (Every 3 Years)	Category II (Every 4 Years)	Category III (Every 5 Years)
Size and Type	Major Manufacturing, Mining or Processing	Minor Manufacturing, Mining or Processing	Warehouses, Real Estate, Administrative Buildings
Employee Safety	Lost Workday Case Incident Rate Worse than Industry Average	Lost Workday Case Incident Rate at Industry Average	Lost Workday Case Incident Rate Better than Industry Average
Process Safety	Covered by the Process Safety Management Rule	Covered by the Process Safety Management Rule	Not Covered by the Process Safety Management Rule
Chemical Exposure	Covered Under >10 1910.1001-50 Chemicals	Covered Under 3-10 1910.1001-50 Chemicals	Covered Under <3 1910.1001-50 Chemicals
Air Emissions	Major source of air toxics or significant emissions; Multiple permits	Moderate emissions; Some air permits	No sources require air permits
Community Relations	Major documented problems with the community	Periodic formal complaints	No or isolated complaints
Hazardous Materials Releases	Has 3 or more §313 chemicals	Has 1 or 2 §313 chemicals	Has no §313 chemicals.
Hazardous Waste	Large Quantity Generator	Small Quantity Generator	Conditionally Exempt Small Quantity Generator
Wastewater	Operates on-site treatment or pre-treatment plant	Discharges Process Wastewater to POTW	Discharges only sanitary wastewater or no discharges
Spill Potential	On-site bulk petroleum or hazardous substances storage of >50,000 gallons	On-site bulk petroleum or hazardous substances storage of 1,000 to 50,000 gallons	On-site bulk petroleum or hazardous substances storage of <1,000 gallons

scoring system. So, a site will not necessarily have to have all of the characteristics associated with a Category I site to be classified as Category I. It may only have one or more of those characteristics to be classified as such. Based on the Coordinators' evaluations, an annual schedule will be published and distributed by the audit program director in December of each year to corporate and group management. Site and group management may request to have any site audited more, but not less, frequently than as determined by the annual program schedule.

3. A Large Construction Company

Audits are scheduled using a formal Risk Ranking tool, which is completed for each site every two years. This aids substantially in prioritizing sites based on risk. Facilities that fall in the high-risk category are audited once every five years. Medium risk sites are audited once every 10 years. Low risk sites are audited if a request is made by the facility, business unit, or law department, or if the site is near high/medium risk sites that are to be audited, or if high and medium risk sites have been audited within the last 5/10-year cycle. Generally, the company relies on the site self-assessment process to address low risk sites.

4. A Medium-sized Chemical Company

The company has developed a site assessment frequency algorithm based on risk. Classes of facilities are assigned frequencies ranging from once every two years to once every 10 years, based on relative risk. Major facilities are generally assigned a frequency of every 2 to 3 years. The audit frequency for a particular facility type is defined based on several criteria, including:

- Relative issue impact/exposure in the operations
- Hazard analysis/risk assessment results
- Prior assessment results
- Accident/incident experience
- Compliance
- Corporate requirements.

5. A Large Chemical Company

EH&S audits shall be conducted at least every three years unless the region program manager extends audit frequency to four years for a particular site or process unit considering the factors described below:

- The existence of an effective first-party EH&S audit program
- Legal or regulatory requirements
- Performance on EH&S metrics and prior audits
- Potential hazards
- Type of site or process unit (e.g., office or warehouse)
- Management-of-change considerations (e.g., turnover of EH&S and management personnel and processes)

6. A Medium-sized Agricultural Products Company

The frequency and scope of the periodic audits will be defined by Corporate EH&S management, and will depend on facility size, complexity, performance information, regulatory compliance history and other appropriate risk factors. The frequency will be documented in a rolling five-year audit plan, which will be reviewed and revised annually by Corporate EH&S management.

7. A Public Power Authority

Audits will be conducted of the Authority's operating projects consistent with the following schedule:

Audits of any Authority facility can be conducted more or less frequently than the above schedule, based on certain risk factors, including, but not limited to:

- Results of the previous audit
- Results of environmental performance metrics
- On-time closure of audit action items
- Extent of change (e.g., people, equipment, regulatory requirements) at the operation

8. A Large Electric Utility

The audit program has established a ranking system to determine the required audit frequency. This system is based on the size and complexity of the site, degree of EH&S risk, history of compliance, financial liability, and prior audits. The major sites are audited approximately once every two to three years, and low priority sites are audited approximately every four or more years. The frequency criteria are adequately defined and communicated and stakeholders agree that the audits occur according to an appropriate schedule.

9. A Major Oil and Gas Company

Audits are to be conducted at the Business Unit level. Audits shall address compliance with the requirements of each process in each subsidiary organization with the following frequencies:

TABLE 10.6
Audit Frequency by Project Type

Project Type	Frequency
Power generation	Once every 3 years
Substations	Once every 4 years
Ancillary operations	Once every 5 years

TABLE 10.7
Audit Frequency by Process Risk

Process Type	Audit Frequency
High risk processes	All organizations to be audited on a three-year cycle
Medium risk processes	All organizations to be audited on a five-year cycle
Low risk processes	Frequency to be designated by responsible organization

The business unit, including each subsidiary, may elect to increase audit frequency.

The design of an audit plan and audit frequencies may take into account any scheduled or completed external audits that adequately address process verification. These external audits could come from regulatory agencies, joint ventures or other partners, or certification bodies.

Note

1. A version of this chapter was previously published by L.B. Cahill in the on-line *EHS Journal*: "Using Risk Factors to Determine EHS Audit Frequency," April 23, 2011.

11

Outsourcing Corporate EH&S Audits

Does it Make Sense?

"No person will make a great business who wants to do it all himself or get all the credit."

—Andrew Carnegie

Introduction

Environmental health and safety (EH&S) audits are an effective compliance assurance tool for many organizations. They are designed to verify whether sites, facilities, and operations are meeting the expectations of applicable regulatory requirements, corporate policies and standards, and good risk management practices. As such, EH&S audits are an important part of corporate governance.

The audit function can take a variety of forms depending on the nature and history of the organization and available resources. These can include:

- A corporate audit function with full-time corporate auditors
- A corporate audit function with part-time auditors selected from business units, divisions, plants, and third-party organizations
- A small corporate oversight group with delegation of the program to the business units or divisions
- A small corporate oversight group and the use of external, independent auditors.

In the past few years there has been substantial debate over whether it makes sense to outsource this function entirely. And, in fact, many, but not all, organizations have done just that. This outsourcing is typically not an easy decision. There are many factors that come into play. This chapter discusses those factors, addressing both the advantages and challenges of outsourcing the corporate EH&S audit program.[1]

Advantages

Outsourcing can achieve any number of important objectives for an EH&S audit program. Among them are:

Readily Available Functional Expertise

In utilizing third parties to conduct the audits, one can typically draw from a large pool of available functional experts and experienced auditors. This is often not the case when using internal staff where the same functional experts are routinely relied upon to conduct the audits. These are staff who, because of their knowledge and expertise, can ill afford to allocate too much of their time to auditing, as they have other organizational responsibilities. There is often substantial tension between the needs of the audit program to staff the audits adequately and functional EH&S managers who need to develop and implement compliance programs. Internal staff is often spread too thin and find themselves in a "no win" situation. Utilizing third-party experts can help to relax this tension.

Local Knowledge

Multi-national companies have the challenge of operating in multiple, if not numerous, jurisdictions. Although English-language international audit protocols are available for purchase, use of these protocols by corporate auditors is sometimes not enough to assure that applicable regulatory requirements are evaluated comprehensively. Further, in countries where English is not the official language, it is advantageous to utilize local third-party auditors, who speak and read the language fluently and who know the customs and mores of the country. Where local auditors are not used, the audit team is often required to hire a translator or to have site staff translate the contents of permits, plans, and other documents during the course of the on-site visit. Neither of these is a very efficient nor illuminating process. As the story goes, there once was this auditor who was interviewing a site environmental coordinator through an interpreter. The auditor asked a fairly direct, short question, which was then translated. The response from the coordinator took about five minutes. At the conclusion of the coordinator's rather extended monologue, the auditor asked eagerly of the translator, "What did he say?" The answer was, "Yes!" Maybe something was lost in the translation.

Greater Independence

Due to a number of events (e.g., Enron and Arthur Andersen) and regulatory responses to those events (Sarbanes-Oxley Act of 2002), the financial auditing profession is under increased scrutiny. This also holds true for EH&S auditing. There is a greater expectation that auditors be truly independent and objective. While the use of third-parties does not meet a full test of independence (e.g., the company can discontinue the use of the third party or refuse payment if the results of an audit are not considered "acceptable"), this approach is generally considered more independent than the use of internal auditors. It can also eliminate conflict-of-interest situations where internal, corporate auditors are asked to audit programs and/or procedures that they helped to develop in their roles as subject-matter experts.

Fresh Set of Eyes

Often, simply having a fresh set of eyes on an audit team can help to identify issues and non-compliances that have been "in the background" for years. Moreover, to use an analogy, the Sarbanes-Oxley makes it *unlawful* for lead third-party financial auditors to provide that service for more than five consecutive years. They are required to sit out an audit at least once every five years. There is a sense in the audit community that auditors can become stale or all too familiar with the operations they are reviewing.

A wonderful example of the need for a fresh set of eyes is depicted in the photograph to the right. This is a site in Australia where a maintenance technician arrived at a unique solution for the disposal of small quantities of waste solvent. He welded a pipe with a funnel onto the roof drain of the building, which of course discharges eventually to a nearby creek. This practice had not been identified on any previous audits until one year a new auditor asked the question, "What is this thing?" It became the "finding of the audit."

Disruption

In most cases where internal staff is used as auditors, these are only temporary assignments. These staff typically have other responsibilities that must be managed almost as an afterthought during the audit week. And if a crisis brings them front and center, then the audit suffers. Either way, this can be disruptive to the EH&S function. Use of third-parties eliminates this issue. The third party auditor will be assigned full time to the audit during the course of the audit.

Challenges

To be fair, outsourcing of an EH&S audit program is not without its challenges. Some of the more significant are highlighted below with thoughts on how to best address them.

Retention of Knowledge

There is a legitimate concern that the knowledge gained by auditors is lost to the organization should third parties be used as auditors. Each audit is a true learning experience, where both non-compliance trends and best management practices are typically identified. Should third parties be utilized, there should be a formal mechanism put in place to assure that strategic observations, which go beyond what is included in the audit reports, are communicated periodically to corporate EH&S management. This might involve a semi-annual or annual meeting with senior management where the following topics are discussed:

- Highlights of liabilities most affecting the corporation
- Trends in the types of non-compliance items to identify potential corporate-wide issues
- Trends in the number of repeat and Level I findings
- Identification of best practices worthy of incorporation by all sites

As an added benefit, this reporting to senior management is very consistent with the expectations of the U.S. Sentencing Commission's November 2004 Guidance on mitigating factors. The guidance states that for a compliance and ethics program to be effective, a company's "governing authority" (i.e., Board of Directors) must be informed, no less than annually, on the status of the company's compliance.

Cost

There is no denying that the use of a third party to conduct audits will result in costs for both auditor labor and travel expenses. However, use of internal staff will result in comparable charges, although the labor costs are not always directly visible to management where part-time auditors are utilized. That is, these internal auditors often don't bill their time directly to the audit; the time is incorrectly viewed as "a free good." In fact, were true cost accounting to be used, an argument can be made that an outsourced program might actually be cheaper. The third party will be expected to maintain trained auditors, updated protocols, and other program tools at their expense. Further, use of a third party can minimize travel expenses, as the intent is to utilize staff in the country or state in which the site is located. This eliminates the need to send corporate EH&S staff half way around the world to conduct audits.

Organizational Nuances

Every organization is unique. And there is no doubt that internal auditors better understand the uniqueness of their company when compared to third parties. Yet, with a certain amount of effort, organizational nuances can be communicated to third-party auditors. Some of the issues that need to be communicated for an EH&S audit program include:

- What is the history of EH&S auditing in the company? Does the program have the organizational clout and top management support it needs?
- Are there corporate EH&S standards and guidelines that must be evaluated on the audits? Which are considered mandatory?
- How will the legal department be involved?
- What can be left behind with the site at the end of the audit? Nothing, a summary of the findings, a draft report?
- How will repeat findings be handled?
- Will findings be classified by significance or priority?
- Will sites be scored? And, if so, how? Is there a pass-fail threshold?

- If there is a disagreement on a finding between the site and the audit team, who will arbitrate?
- What are the records retention and destruction policies for audit working papers and reports?

Attention to these details up front will make for a better program and a more informed and savvy third-party auditor.

Familiarity with Operations

One of the principal concerns leveled at third party auditors is that they are not always sufficiently familiar with the operations of a given company or industry to understand the subtleties involved in achieving compliance. Mining is different from power production, which is different from pharmaceutical manufacturing, and so forth. To some extent, there is a certain amount of truth to this concern. On the other hand, quite often a lack of full familiarity results in auditors asking that one "naïve" question (e.g., "Why exactly is that sump not considered a confined space?"), which strikes at the heart of an issue not previously identified. In any event, there are two solutions to this challenge. One, third party auditors can be taught about the unique aspects of a business. It is something that can be learned. Or two, the pool of third-party auditors is very large; people with knowledge of a particular business can almost always be identified.

Why Not a Blend?

One solution used by numerous organizations is to design and implement a program that utilizes a blend of internal and external resources. That is, internal auditors are used where feasible, and they are supplemented by third-party auditors where local knowledge and presence or a particular expertise (e.g., process safety management) is needed. This approach has some very distinct technical and cost advantages. It also adds a layer of independence that would not exist where only internal resources are utilized.

Note

1. A version of this chapter was previously published by L.B. Cahill in the on-line *EHS Journal* as: "Outsourcing EHS Audits: Does it Make Sense," November 17, 2010.

12

The Use of Statistically Representative Sampling on EH&S Audits

Does It Apply?

"Prediction is very difficult, especially if it's about the future."

—Niels Bohr

Sampling is a common technique used during EH&S audits when a large universe of items needs to be reviewed. Statistical representative sampling in particular can provide a means to identifying the number of items to be reviewed to be fairly confident the results are representative of the entire set of data. Although considered an accepted practice in auditing, statistically representative sampling is undertaken only occasionally, at best. Sampling is typically conducted in a less rigorous fashion. In that context, this chapter discusses the following three topics:[1]

- Sampling expectations of third parties
- Using statistically representative sampling on audits
- An alternative method of sampling: the 10% rule.

That is, the chapter answers the questions: what is expected by third parties of an audit program with respect to sampling; is it feasible to expect statistically representative sampling to meet those expectations; and, if not, what might be an alternative approach that would stand up to scrutiny?

Sampling Expectations of Third Parties

About five years ago OSHA conducted a Process Safety Management (29 CFR 1910.119) inspection at BP's Husky Refinery near Toledo, OH. The inspection resulted in 42 willful violations and approximately $3 million in proposed penalties. One of the actions proposed by OSHA was that BP "audit a *statistically significant* number of pressure vessels, piping and instrument controls during the company's PSM compliance audits." An article[2] published by the author in 2010 reported that this was the first time that statistically representative sampling had been proposed formally as part of any EH&S audit program.

In a comment on the article recorded in the *EHS Journal*, Joel Olener stated that this actually was not the first time this had happened with OSHA under the PSM standard. There was a previous case where OSHA had cited a company for not using statistical sampling for the audit. After reviewing the PSM standard Mr. Olener reported that the only place with any reference to "sampling" was the NON-MANDATORY (emphasis added) Appendix C quoted below:

> An audit is a technique used to gather sufficient facts and information, including statistical information, to verify compliance with standards. Auditors should select as part of their preplanning a sample size sufficient to give a degree of confidence that the audit reflects the level of compliance with the standard. The audit team, through this systematic analysis, should document areas which require corrective action as well as those areas where the process safety management system is effective and working in an effective manner. This provides a record of the audit procedures and findings, and serves as a baseline of operation data for future audits. It will assist future auditors in determining changes or trends from previous audits.

What these OSHA initiatives do not do is define the term "statistically significant." This opened up the question as to what other third parties expect with respect to statistically representative sampling. As a consequence, the following third-party auditing standards were evaluated for statements related to sampling on audits:

- The Board of Environmental, Health and Safety Auditing Certifications Standards, 2008
- ISO 19011 Auditing Guidelines, 2002
- Auditing Roundtable Standards, 1993
- U.S. EPA Auditing Policy, 1986, 2000
- Institute of Internal Auditors Standards, 1997

A review of the above standards strongly suggests that only "appropriate sampling" or something comparable is the recommended practice. See the statements directly below for the standards' language that applies to the use of sampling techniques on EH&S audits.

BEAC Standards

"When conducting an audit, EH&S auditors shall use due care in examining and evaluating information they gather. This information shall be sufficient, complete, relevant, and useful to provide a sound basis for audit findings and recommendations." (Section III.8)

"Performance and Program Standards for the Professional Practice of Environmental, Health and Safety Auditing," Board of Environmental, Health and Safety Auditor Certifications (BEAC), 2008.

ISO 19011 Auditing Guidelines

"Audit evidence is verifiable. It is based on samples of the information available, since an audit is conducted during a finite period of time and with finite resources. The appropriate use of sampling is closely related to the confidence that can be placed in the audit conclusions." (Section 4[e])

"During the audit, information relevant to the audit objectives, scope and criteria, including information relating to interfaces between functions, activities, and processes, should be collected by appropriate sampling and should be verified. Only information that is verifiable may be audit evidence. Audit evidence should be recorded. The audit evidence is based on samples of the available information. Therefore there is an element of uncertainty in auditing, and those acting upon the audit conclusions should be aware of this uncertainty." (Section 6.5.4)

"Guidelines for quality and/or environmental management systems auditing," International Organization for Standardization, ISO 19011:2002(E)

Auditing Roundtable Standards

"While on site, auditors must gather information necessary to fulfill the audit objectives. The information collected must be relevant, accurate, and sufficient to support findings, conclusions, and recommendations. Appropriate sampling schemes should be utilized in selecting samples." Section II (C) (3)

"Minimum Criteria for the Conduct of EHS Audits," The Auditing Roundtable, 1993

U.S. EPA Auditing Policy

"A process which collects, analyzes, interprets, and documents information sufficient to achieve audit objectives." (Appendix, Section V)

"Environmental Auditing Policy Statement," US Environmental Protection Agency, July 9, 1986

Institute of Internal Auditors Standards

"Information should be sufficient, competent, relevant, and useful to provide a sound basis for audit findings and recommendations.

a. Sufficient information is factual, adequate, and convincing so that a prudent, informed person would reach the same conclusions as the auditor.
b. Competent information is reliable and the best attainable through the use of appropriate audit techniques.
c. Relevant information supports audit findings and recommendations and is consistent with the objectives of the audit.
d. Useful information helps the organization meet its goals.

Audit procedures, including the testing and sampling techniques employed, should be selected in advance, where practicable, and expanded or altered if circumstances warrant." (Sections 420.2 and 420.3)

Standards for the Professional Practice of Internal Auditing, The Institute of Internal Auditors, 1997

Using Statistically Representative Sampling on Audits

Since sampling is obviously an integral part of any audit, should audit program managers attempt to add rigor to the process by making the sampling statistically representative? In an ideal world, the answer would be yes, but let's explore what this might imply. Sampling strategies typically involve three steps:

- Define the universe or population size to be sampled (e.g., employees, portable fire extinguishers, above ground storage tanks)
- Decide on the sample size (i.e., how many should be sampled)
- Determine the sampling method to be utilized (e.g., random, systematic, stratified, cluster)

Estimating population size should be a fairly straightforward exercise in most cases. For example, the site should be able to determine the total number of:

- Employees, through human resource records
- Portable fire extinguishers, through the on-site inspection program
- Wastewater discharge monitoring reports, typically a monthly requirement
- Hazardous waste accumulation area inspections, typically required weekly
- Aboveground storage tanks, through observation if nothing else
- Safety data sheets, taken from the chemical inventory.

In order to determine a sample size that is statistically representative, statistical theory must be employed. Without belaboring the mathematics, the approach is detailed in Table 12.1. Completing the equations in Table 12.1 results in the estimates in Table 12.2. That is, sample sizes for various populations have been estimated given one of three confidence levels (85%, 90%, and 95%). The results using other confidence levels can be calculated using the equation in Table 12.1.

One of the most important assumptions that needs to be made in conducting sampling is what confidence level or interval is acceptable. A confidence level is the probability that the results of the sampling will be truly representative of the population. In general practice, most statisticians presume that a minimum confidence level of 95% is required.

Let's see what the above means by way of an example. Assume that there are 50 machines or pieces of equipment on site that require guarding. The auditor wants to determine if the guards are in place and in good repair. At a 95% confidence level, how many machines would the auditor have to inspect in order to make a determination? Using Table 12.2, the answer would be 44.5, or 45 (89% of 50 from column A). Of course, one would ask the question, if I have to look at 45, why don't I just look at all 50? And there's the rub. With the typically small population sizes of EH&S activities, it often does not make sense to conduct statistically representative sampling given the short amount of time auditors spend on site.

And lastly, once the sample size has been selected (regardless of whether statistically representative sampling is utilized), then the sampling method must be chosen. Common methods include:[3]

- **Simple Random Sampling**. A sample in which every member of the population has an equal chance of being chosen. A standard random number generator can aid in this

TABLE 12.1
Sampling Theory

Sampling theory can provide a means to identifying the number of items to be reviewed to be fairly confident that the results are statistically representative. That is, when a confidence interval of a given width is desired for an unknown proportion, p (in this case the proportion of sources that are in compliance), the sample size, n, required can be obtained using a formula given by Inman and Conover.[a] However, the formula applies to an infinite population. A correction for a finite population is given by Cochran.[b] The finite population correction is close to unity when the sampling fraction n/N remains low (approximately 5% or less). Combining the two formulae gives:

$$n = (4z[\alpha/2]^2 pqN)/(w^2N - w^2 + 4z[\alpha/2]^2 pq)$$

where:
n is the required number of sources in the sample
$z[\alpha/2]$ is the $(1 - \alpha/2)$ quantile from the standard normal probability function, that is:

Confidence Level	α	$1 - \alpha/2$	$z[\alpha/2]$
95%	0.05	0.975	1.96
90%	0.10	0.950	1.65
85%	0.15	0.925	1.44

p is the proportion of interest (proportion of sources in compliance); when p is unknown it is set equal to 0.5 to give the most conservative possible value of n
q is $(1 - p)$
N is the total population being sampled
w is 2α

[a]Inman, R.L. and Conover, W.J., *A Modern Approach to Statistics,* John Wiley and Sons, New York, NY, 1983
[b]Cochran, W.G., *Sampling Techniques,* Second Edition, John Wiley and Sons, New York, NY, 1977

selection process. An example would be to randomly sample from an entire five-year file of hazardous waste manifests.

- **Systematic Sampling.** A sample that involves randomly listing the population in order and them picking every tenth, hundredth, or thousandth, and so on, item from the list. An example would be to sample every tenth hazardous waste manifest over a five year period.
- **Cluster Sampling.** A simple random sample of groups, or clusters, of the population. Each member of the chosen clusters would be part of the final sample. An example would be to choose all manifests for a particular vendor where historical problems have been identified.
- **Stratified Sampling.** Sampling obtained by dividing the population into mutually exclusive groups, or strata, and randomly sampling from each of those groups. An example would be to randomly select manifests from each of the three waste vendors being used by the site.

Where sampling techniques are utilized on an audit, teams should be careful to document the approach taken. In writing findings, providing three specific numbers is especially important in order to provide perspective: the total number in the population, the number sampled, and the number of deficiencies. Thus, an example finding might state: "There were 250 portable fire

TABLE 12.2
Selecting a Sample Size on EH&S Audits*

Population Size	Suggested Minimum Sample Size		
	A	B	C
10	98%	88%	72%
25	94%	74%	49%
50	89%	58%	32%
100	80%	41%	19%
250	61%	21%	8%
500	43%	12%	4%
1000	28%	6%	2%
2000	16%	3%	1%

A—Suggested minimum sample size for a population(s) being reviewed which is considered to be extremely important in terms of verifying compliance with applicable requirements, and/or is of critical concern to the corporation in terms of potential or actual impacts associated with non-compliance. A confidence interval of 95% is assumed.

B—Suggested minimum sample size for a population(s) being reviewed that will provide additional information to substantiate compliance or non-compliance and/or is of considerable importance to the corporation in terms of potential or actual impacts associated with non-compliance. A confidence interval of 90% is assumed.

C—Suggested minimum sample size for a population(s) being reviewed that will provide ancillary information in terms of verifying compliance with a requirement. A confidence interval of 85% is assumed.

Source: Inman, R. L. and Conover, W. J., *A Modern Approach to Statistics,* 1983; and Cochran, W. G., *Sampling Techniques, Second Edition,* 1977, John Wiley & Sons, Inc., New York, NY.

extinguishers at the site. Twenty were inspected by the audit team, and four were determined to not have had the latest monthly inspection recorded."

An Alternative Method of Sampling: The 10% Rule

What can be done as an alternative to conducting statistically representative sampling? Within the profession, there is something called the 10% rule. This means that when in doubt, a good alternative estimate would be sampling 10% of the population. This works quite well in many cases. However, for very small and very large populations, it doesn't work very well. For example, if there are only 10 aboveground storage tanks on site, inspecting only one seems insufficient at face value. Conversely, if there are 1,000 leak detection and repair (LDAR) sources that are inspected and monitored by the site, having the auditor inspect 100 of these might be considered overkill. So, in an effort to modify the 10% rule, Table 12.3 was developed. This is only a rule of thumb and has little if any scientific foundation. Still, the 10% rule can be quite helpful when statistically representative sampling does not make sense.

TABLE 12.3
An Alternative: The 10% Rule

Population Size	Percentage Sampled
1–50	>10%
51–500	10%
>500	<10%

Conclusion

The audit initiative by OSHA, a small part of a very significant enforcement action, could mean the beginnings of more rigorous sampling expectations on EH&S audits. Or not. The best estimate is that, due principally to limited time and resources, presently very few, if any, companies use formal statistically representative sampling on EH&S compliance audits. This might change over time for any number of reasons.

Notes

1. A version of this chapter was previously published by L.B. Cahill in the online *EHS Journal* as: "Statistically Representative Sampling on EHS Audits: Expectations," Established by Third Parties, June 21, 2010. This chapter has been revised and expanded significantly from the original article.

2. Ibid.

3. Adapted from Donnelly, Jr., PhD, Robert A., *The Complete Idiot's Guide to Statistics*, The Penguin Group, 2007, pp. 165–176.

13

A Final Perspective

Lessons Learned in 35 Years of Auditing

"Learning never exhausts the mind."

—Leonardo da Vinci

Introduction

I started my EH&S audit career back in the late 1970s, some 35 years ago as of this writing. My first significant involvement was as a consultant, working with my beloved mentor, Dr. Myron Weinberg, a Senior Vice President at Booz, Allen & Hamilton. We were at a Dallas-Fort Worth Airport meeting with senior EH&S managers at a Fortune 100 chemical company. The purpose of the meeting was to convince our client that we could build audit protocols for them as they were embarking on the development of a full-scope environmental audit program. At the time, commercial protocols did not exist. Using some classic consulting techniques (i.e., stretching the truth a bit) we convinced the client that we could hit the ground running because of our knowledge of the topic and some "ready-made" materials. While this was partly true, our proposed cost estimate for the project suggested that we were much further along than we actually were. We won the project (the good news), and I was designated the project manager who had to create the deliverables for the proposed cost (the bad news). And that was my first substantial exposure to auditing.

Some 15 years later in 1995, I was working with my friend and colleague Lori Michelin who was then the Environmental and Occupational Health and Safety (EOHS) Audit Program Manager, and is now Vice President-EOHS and Product Sustainability at Colgate-Palmolive. We were training potential CP EOHS staff on how to conduct audits. The training occurred in Malaysia and included two initial days of classroom training and a subsequent three-day site audit as part of the process. As Lori and I observed our just-trained auditors on their first audit we noted that for many of them the training had only sunk in partially. So, during the field audit, we developed a list of 20 suggestions for the new auditor, which we used to sum up the lessons that should have been learned during the training. This magical list was the basis of an article later published in 1996 in Wiley's *Environmental Quality Journal* entitled "Achieving Quality EHS Audits: Twenty Tips for Success."[1]

It is now almost 20 years since the publication of the "Twenty Tips" article, and I took time to look back and see if the advice still holds up. Mostly it does, but, not surprisingly, I believe that I have learned even more about the audit process in the past 20 years. Herein is my final list of ten lessons learned and to be learned on EH&S audits. Where it's meaningful, anecdotes are provided to support the lesson.

1. Listen, Listen, Listen

In the past few years it has been fascinating for me to be in a position to observe audits as a third-party evaluator. One of the more interesting aspects to observe has been the interviews between auditors and site operators. There is a long-standing rule that during an interview, the interviewer should be talking 20% of the time and the interviewee 80% of the time. In practice, all too often, it is just the opposite on audits. Auditors become bogged down with their audit protocols and the ten-thousand questions contained therein. So the process becomes a rattling-off of scores of closed-ended questions by the auditor and cryptic yes/no answers by the site staff. Auditors need to stick to the 80/20 rule and let the operator talk freely and openly – listen, listen, listen. After all, the word "audit" is derived from the Latin word "audire," which means "to listen." I must admit that I have learned this the hard way myself as evidenced by a couple of auditor training experiences that I have had. Communications, especially in international settings, can be quite challenging.

Auditor Training in France. Some years ago I was asked by a mid-sized chemical company to lead a training program for new auditors. The training session was held in rural France and included people from all over the continent. About a half hour into the training, a student raised her hand, and I'm thinking, "Wow, I'm really getting through to these people: a question already." Well, the question was "Monsieur Cahill, could you please hurry up and slow down?" Say what, I thought! It turned out that what she meant was that the material was very interesting, and she wanted me to make sure that I got through it all but that I was speaking too fast for her to comprehend it all in her second language. So, a bit perplexed, I tried to be more articulate with my speech, slow it down a bit, but plow on through. A lesson learned in international communications for me.

Auditor Training in Brazil. I was leading a new auditor training class for a large consumer products company in Sao Paulo, Brazil. The course, held for South American EH&S staff, included two and a half days of classroom training followed by a two and a half day field audit of the site where the training was held. The training seemed to go quite well as there was a lot of interaction. In fact, one student, Martin from Argentina, came up to me after the class to tell me in broken English how terrific the experience was for him personally. He even wanted an autographed copy of my book. He was effusive in his praise; my head swelled. Later when reviewing the very positive course evaluations, which ranked each of the course elements from 1 to 5, 5 being outstanding, I noticed there was one completed evaluation unlike any of the others. It had scored each element a "1," or the worst score possible. I felt I had to speak to this individual right away to find out why they were so unhappy with the course. And sure enough, it turned out that the form was signed by Martin of all people. Sadly, he had already left, and I never did find out what the problem was. My assumption was that I was not clear in discussing how the evaluation forms were constructed and that Martin's English was so poor he simply assumed that "1" meant the best. Well, at least that's my story, and I'm sticking to it!

2. Do Your Homework

Auditors really need to come armed with a knowledge of applicable requirements when they visit a site to conduct an audit. This has become much easier since the development of commercial U.S. and international audit protocols. Yet, the protocols often do not get at the subtlety of some of the federal and state regulations. Moreover, local regulations and permits themselves typically are not covered by the protocols. So, much homework needs to be done prior to the audit. And this needs to be done well in advance not on the plane ride to the site. For those who have a full-time job other than auditing, which the majority of us do, this preparation and pre-audit planning can be quite a challenge. However, it must be done, otherwise problems can occur.

Hazard Communication Auditing in North Carolina. Our team was conducting an EH&S audit in North Carolina. In reviewing the audit protocol for Hazard Communication, the H&S auditor noted that the training provisions of the regulations (29CFR1910.1200(h)(2)), "require that employees *shall be informed* of the location and availability of the written hazard communication (HazCom) program including the required list(s) of hazardous chemicals, and MSDSs required by this section." In interviewing four employees, each was able to identify where the nearest MSDSs were located. Moreover, when requested, each was able to locate a specific MSDS for a fairly obscure water treatment chemical, and they knew how to "read" the MSDS. However, none of the four employees were able to tell the team where they might find the written hazard communication program. So did we have a finding? The team's ultimate conclusion was no. We finally did verify that the employees had received HazCom training during which the location of the written program was communicated; they just didn't remember that part of the training some years later. Because the staff were otherwise well informed and capable, cooler heads prevailed.

Stormwater Auditing in Texas. About a year ago, I was asked to review an audit report for an audit done at a Texas power plant. It seemed that there was a disagreement between the auditors and the site over the results of some stormwater sampling. Samples had been taken during a rainfall event and there were exceedances for lead and zinc. The auditor wrote this up as a finding; a violation of the Texas General Stormwater Permit. The site, on the other hand, claimed that the samples were done for benchmarking purposes and therefore were not violations. The auditor retorted back that exceedances from any sampling event are violations and must be reported as such. And on and on it went. What no one thought to do was to take a detailed look at the actual permit. And the language below finally clarified the matter:

> The permittee shall compare the results of analyses to the benchmark values listed below in Table 3 for any pollutant(s) that the permittee is required to monitor in this general permit, and shall include this comparison in the overall assessment of the SWP3's effectiveness. *Analytical results that exceed a benchmark value are not a violation of this permit,* as these values are not numeric effluent limitations. Results of analyses are indicators that modifications of the SWP3 may be necessary.[2]

In this case, the auditor was found to be incorrect in his conclusion; the consequence of not doing the homework required.

3. Be There When Things are Happening

Too often on audits the audit team is escorted into a conference room for the opening meeting, then given a windshield tour of the facility and returned to the conference room, virtually never to be seen again. This is a bit of an overstatement but has been actually observed on some audits. Unfortunately, there is a tendency for audit teams to get bogged down in the paperwork and to not observe activities at the site as they are happening. This is especially true for environmental (versus health and safety) auditors as they deal with air and water permits, monitoring records, hazardous waste manifests, training records, inspection logs, and the like.

This can be a real problem if the goal is to determine the site's compliance with requirements. For example:

A Distribution Audit in Bulle, Switzerland. I observed a distribution or transportation audit of a medium-sized chemical plant in Bulle, Switzerland, a city located in a region famous for Gruyère cheese. The site was the largest user of liquid nitrogen in the country. They received two full tanker trucks per day of nitrogen. The hookup for the tanker truck was actually located outside the fence line and had very little security associated with it. During the course of the three and a half day audit, the team did not observe a single delivery of nitrogen! They simply were overwhelmed with the paperwork and, after the initial site tour, spent pretty much the entire time in a conference room reviewing records and interviewing staff. They did observe one truck leaving the site with finished product in drums, boxes, and containers. Even the plant manager was surprised, if not thankful, that the team did not observe one liquid nitrogen delivery during the audit.

There are many potential activities that could and should be observed on EH&S audits. Candidates include:

- Safety Meetings
- New Construction or Demolition
- Emergency Response Drills
- Inspections and Sampling
- Loading and Unloading
- Materials Handling and Transfer
- Process and Equipment Startups or Shutdowns
- Waste Packing and Pickup
- Confined Space Entries
- Lock, Tag, and Try

One of the first tasks for an audit team when they arrive on site is to ask if they could observe some of the above activities during the audit. But be forewarned. This can lead to some interesting responses from the site, such as: "Sorry, we had five confined space entries last week but, sadly, none are scheduled for this week." or "Interesting you asked to see that, but we had a major hazardous waste pickup just this last Friday. There are none scheduled for this week." Alternatively, you can have just the opposite response when requesting to observe an activity as described directly below.

Emergency Response Drill on an Environmental Audit in Venezuela. I led an environmental audit of a large chemical plant in Venezuela a few years ago. Being the diligent team leader, I requested in advance that an emergency response drill be conducted while the team was there. And sure enough, a drill was scheduled and conducted. Unfortunately, it was the drill to end all drills. The site simulated a release of oleum or fuming sulfuric acid. The local township was involved, and I saw more emergency response gear and equipment than I have ever seen in my life. The drill took an entire day to complete. During this entire time, the audit team stood behind a taped-off area and observed the process. While this response was fascinating and extremely well done, observing this activity for an entire day was not the best use of our time. And maybe the site knew that? The bottom line is that the team should observe critical site activities, but make sure it's the best use of your time.

4. Understand the Rules of the Road

Almost every audit assesses an operation's compliance with federal, state or regional, and local regulations. However, for many major corporations, an evaluation of performance against company policies is also required. These policies can include both mandatory standards and optional guidelines. This structure often poses challenges for both internal and third-party auditors. For internal auditors, they must be careful to understand that what is adopted at one site as a procedure based on a corporate policy, especially when it's their own site where a practice is adopted, might not be required at all sites. They can't force what might make sense to them based on their personal experiences onto another site during an audit. For third-party auditors, they must become intimately familiar with the client's EH&S policies and truly understand what is considered mandatory and what is not.

Audit programs often require that findings be classified in two ways: by type and by priority. An example classification system is provided in Table 13.1. Auditors must become familiar with any proposed classification system and understand the nuances presented by these. For example, a high priority, Level I finding to one person might be a Level II to another. Calibration among these levels is critical to achieving an agreement among auditors and site management. It should be noted that some companies do not use any form of a findings priority scheme. The presumption is that all findings are equally important and the deficiency must be corrected in a timely fashion. Personally, I am a big proponent of priority classification systems. I have found that many site managers are supporters as well; they want to know what's really important.

The case study presented below, based on an actual experience, demonstrates how corporate standards can come into play in a significant way on an audit.

Foaming Wastewater Discharge in Sao Paulo, Brazil. On an audit of a soap plant in Brazil there was an observable wastewater discharge with significant foaming. Figure 13.1 is a photograph of the discharge at the final effluent box. This discharge continued down to a local stream where foam was also visible. The foaming was due to residual surfactant, a product component that purposely creates the foam in soap. This particular surfactant was quite powerful and existed in the effluent at a concentration of 3 mg/l. The site's permit allowed for a concentration of 5 mg/l. So the plant manager stated, correctly, that he was in compliance and wanted to know

TABLE 13.1
Classification of EH&S Audit Findings—Sample Definitions

Type of Finding

REGULATORY: A finding involving laws, ordinances, regulations, directives, etc. that are external to the company. These include government regulations and international treaties. Regulatory findings may include regulatory guidance that is deemed to be enforceable or provisions/procedures included in required plans or programs (e.g., SPCC Plan, Confined Space Entry Program).

POLICY: A finding involving corporate, business unit, division, or site policies or mandatory standards.

GUIDANCE/OTHER/COMMENT: A finding involving non-mandatory corporate guidelines, practices, or systemic controls that may favorably impact EH&S performance if implemented at the site.

Finding Priority

LEVEL I (RED): HIGHEST PRIORITY ACTION REQUIRED: Situations that could result in substantial risk to the environment, the public, employees, stockholders, customers, the company or its reputation, or in criminal or civil liability for knowing violations.

LEVEL II (YELLOW): PRIORITY ACTION REQUIRED: Does not meet the criteria for Level I but is more than an isolated or occasional situation. Should not continue beyond the short term.

LEVEL III (GREEN): ACTION REQUIRED: Findings may be administrative in nature or involve an isolated or occasional situation.

FIGURE 13.1
Foaming Wastewater Discharge in Sao Paulo, Brazil

from the auditors what all the fuss was about. He further noted that the community was quite pleased with the discharge. It seemed that the nearby residents were generally quite poor and they would come to the stream and wash their clothes below the discharge to take advantage of the residual soap and would rinse their clothes above the discharge, where the water was clear. This was a "win-win" as far as the plant manager was concerned. Unfortunately for him, this company had a mandatory corporate standard that prohibited any visible foam in any wastewater discharge. And this applied anywhere they operated throughout the world; no exceptions. The result, of course, was that this foaming issue was an audit finding, and the situation had to be corrected. In some ways, this didn't seem fair but the auditors' hands were tied. If the company wanted to make an exception in this case that would be left to senior line executives.

5. Be Wary of Just-in-Time Compliance

Do plant managers prepare for an audit by performing a sort of facility triage just prior to the audit? Well, of course they do and there is absolutely nothing wrong with that. At the same time auditors need to be wary of any pre-audit quick fixes implemented at the site that will not stand the test of time. There is a risk that these new processes and procedures will break down once the audit is completed. Examples of how this just-in-time compliance can affect an audit program are presented below.

A Confined Space Entry Audit. On an audit some years ago in Europe at a major chemical plant owned by a U.S. corporation, my audit team was required to review the site's confined space entry (CSE) program. The applicable regulations and corporate standards mimicked OSHA's requirements. A copy of the site's procedure or written program was requested, which was given to the team without hesitation. The procedure tracked the OSHA regulations exactly.

However, the effective date of the procedure was three days prior to the audit now being conducted. The EH&S Coordinator admitted that the procedure was brand new and was completed in anticipation of the audit. When asked about the required training, the coordinator stated that it was scheduled for the following week; they had to wait until issuance of the final procedure and could not do the training during the audit week. There was a fully documented training plan for next week. The team also asked to review any completed entry permits and were told that, since the procedure is so new, no completed permits were available. There were some old completed permits in the file that were used under a previous more informal system. Upon review of a sample of the permits, they were deemed complete but did not address all of the elements of a safe entry. The bottom line was that the site staff seemed to be knowledgeable of the requirements and were in the process of putting an excellent program in place. It was just not fully implemented. What to do, what to do? The team ultimately decided that the audit could not properly assess the confined space entry program and recommended that a focused CSE follow-up audit be conducted in three months. Not a perfect solution, but it was deemed better than a scathing write-up in the audit report.

Construction of Secondary Containment during an Audit. I was a member of an audit team at a large chemical plant in the United Kingdom. We were assessing the above ground storage tanks (ASTs) and loading/unloading (L/UL) areas during the audit. Corporate standards required that all ASTs and L/UL areas have secondary containment sufficient to contain the contents of the tanks or trucks with a 10% additional allowance for rainfall. The site's performance against these requirements was excellent overall. However, there was an ethanol unloading area that did not have secondary containment. This was identified on the very first day and it was clearly going to be an audit finding. The plant manager was made aware of this and immediately began a project to build secondary containment for the area. Figure 13.2 shows it under construction. The project was completed on the last day of the audit and the site even had a formal ribbon-cutting ceremony with the audit team to celebrate. What to do, what to do? It appeared to the team that there were four options:

FIGURE 13.2
Construction of Secondary Containment during an Audit

- Write the situation up as a commendation, an example of a best practice.
- Do not include it in the report; it's a completed action, no follow up is required.
- Write it up as a finding in the report but indicate that the corrective action was completed during the audit.
- Write it up as a finding in the report and let the site indicate what action was taken in the first corrective action report.

The plant manager wanted the team to select the first or second option of course. The hard-core auditors preferred the last option. The third option was finally selected by the team leader as the best approach.

6. Recognize That the Site Staff Has Goals for the Audit

In Chapter 7 it was suggested that one of the stated goals for auditors should be: "We're there to help the site improve." and not: "We're there to find problems. It's almost like we're getting paid by the finding." Auditors should recognize that site management almost always has a goal for the audits as well. A worthwhile goal might be: "We want to use this audit to make us the best we can be." Unfortunately, the goal is, more often than not, something like: "We will have zero (or less than five) findings on this audit, and all staff will be held accountable for meeting that goal." These types of goals are almost never communicated to the auditors but they do exist. To suggest otherwise, as some corporate audit program managers have done to me, is naïve. This setting of goals by the site can have some interesting consequences for the process.

An EH&S Audit of a Large Manufacturing Plant in Europe. On this audit our team was working with a very professional and accommodating EH&S coordinator. The process was smooth over the first day and a half. At the end of the second day when the team began to report its few initial findings in the daily debrief, the coordinator was uncharacteristically a little distant as the meeting ended. On the third day the coordinator was back to "his ol' self" so the team assumed it was just a bad afternoon for him. As the third day progressed and more issues were identified the coolness returned. By the end of the day, the coordinator's behavior was bordering on downright hostile. What to do, what to do? Finally, I pulled him aside and asked him what was going on in the most supportive way I could. It turned out that the plant manager had set a goal of five or fewer findings for the audit and, if there were any more than that, there would be serious consequences. And moreover, as in many plants, the EH&S Coordinator, fairly or not, would experience most of the impact. This kind of pressure would affect anyone's behavior. We weathered through the storm, and the team was careful to communicate the good with the bad during the closing meeting. The site's performance overall was excellent and the fact that there were more than five findings should not have obfuscated that fact.

7. Look for Root Causes; Permanent Solutions

Too often on EH&S audits, the auditors focus on administrative deficiencies, document those, and move along. It's a pretty straightforward exercise to identify three missing inspections, four unsigned manifests, five missing MSDSs, two unlabeled drums, or six wastewater discharge exceedances. It's a more difficult proposition to identify why those situations occurred. Auditors can add real value on audits by digging deeper into any given process to identify possible root causes that might be causing non-compliance. These will typically be related to a breakdown in management systems or process controls.

It is clear, however, that there is not near enough time on a given audit to undertake a detailed root cause analysis for each and every finding. What the auditors can do is to prioritize findings and explore root causes with the site staff for those findings with potential severe consequences. If that process cannot be completed during the course of the audit, then the audit report should require a root cause analysis to be completed by the site as part of the corrective action plan.

This call for conducting more root cause analysis and focusing more on management systems and process controls can add a burden to the audit team. It also will push them to conduct a type of analysis with which they are not comfortable nor do they feel sufficiently trained to conduct. Yet, if conducted in certain critical situations it can add real value to the audit.

Finally, one of the things that I have observed over the years, and am convinced is true, is that a good 90% of all findings of deficiency are due to one of two underlying causes: poor management of change or poor communications. It is my belief that if there are no changes to a particular operation, it can eventually be brought into and kept in compliance. The problem is that this never happens. People change, products change, equipment changes, regulations change, everything changes. And when many of those changes occur, they are not communicated effectively to those who have a need to know. Auditors should ask one key question during the opening conference on an audit: "What has changed since the last audit?"

The Blocked Fire Extinguisher. On the first day of an audit in the U.S., the team came across a situation where a portable fire extinguisher and an emergency eyewash station were both blocked by a battery charger (see Figure 13.3). The site escort was pretty much embarrassed when the situation was pointed out to him. He immediately moved the battery charger to another location. His response was: "I don't know what happened. The guys know better. It's never like this. It must have happened last night." Ah, the "it must have happened last night" excuse. The team believed that the site staff understood the issue; they made an immediate correction; and maybe this was a one-off situation that's been corrected. Thus, it was concluded that this situation would not be listed as a finding in the audit report. At least that was the conclusion until the next day when the charger had returned back to "where it

FIGURE 13.3
Blocked Fire Extinguisher and Eyewash Station

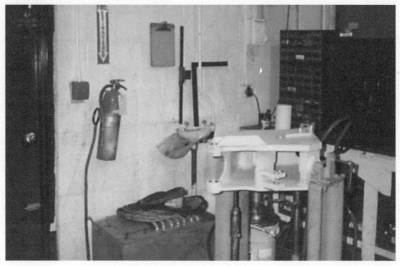

belonged." This clearly was a classic auditor's trap. While the site had addressed the symptom, they had not addressed the root cause of the problem. Why was the charger being placed where it doesn't belong? What kind of controls were in place to keep this from happening and why weren't they working? The team reopened the investigation and explored the issue more deeply. Lesson learned.

8. Know Materiality Thresholds

Deciding what deficiencies are material on an audit and should be included in an audit report as findings is often not a straightforward exercise. For example, what if an auditor knows that there are 500 portable fire extinguishers on a site, 50 are inspected by the team, and only one does not have an up-to-date monthly inspection tag. The fire extinguisher program otherwise is quite sound. Extinguishers are charged; they are accessible; annual training is conducted; and so forth. Most auditors would probably not write this observation up as a finding. In fact, many companies have developed a "local attention only" classification for just this case. These are considered one-off deficiencies that don't meet materiality thresholds and are not included in the audit report but are provided to the site manager at the end of the audit as a list of items that need to be addressed. To me this makes eminent sense. The resultant audit report will not be cluttered with minor deficiencies that create background noise and obfuscate the real issues that need to be addressed. The problem, however, is defining the cutoff for material-ity. If one occurrence does not meet the threshold, what about two, what about three? Where does an auditor draw the line? There is no perfect answer to this dilemma, but a good rule of thumb is to use the "rule of three." Before an issue becomes an audit finding, there needs to be a minimum of three occurrences. The two cases discussed below help to present the quandaries associated with defining materiality.

Auditing MSDSs. On one particular audit, I had the responsibility to review the site's Hazard Communication program. I had just completed my review of the Material Safety Data Sheet (MSDS) file. MSDSs were available for all 200 or so chemicals used at the site and they were generally up-to-date. There was one exception, though. The MSDS for a biocide used only occa-sionally in treating boiler water was not available. There was one 30-gallon container of this chemical stored in the boiler house. The site obtained the MSDS during the audit and placed it in the file. I have come to find out since then that water-treatment chemical vendors are con-stantly changing and improving the chemicals they use as new ones come on the market. As a consequence, the chemicals used in water treatment are actually good candidates for auditors to test the MSDS management system. I have identified problems and deficiencies in this area on roughly 30% of the audits that I have conducted.

The question here, of course, was this a finding? It was only one of 200 MSDSs required to be available at the site and it was retrieved during the audit. Moreover, OSHA regulations require that MSDSs be "readily available" to site staff. In this age of the Internet, the definition for "readily available" is open to interpretation and, in my experience, has caused much discus-sion on audits where an MSDS for a certain chemical is missing from the paper file, as in this case. The conclusion I reached was that this did not reach the required level of materiality to be considered a finding. It was listed as a local attention only item. The risk in doing so was - what if OSHA comes in next week and finds a similar situation? Will management wonder about the effectiveness of the audit? Who will management go to first if there is a fine? The answer to that question is pretty obvious.

Calibrating Interviews. As an audit team member on a large industrial site in England, I was responsible for assessing the emergency response program, particularly employees' knowledge of emergency evacuation procedures. The site selected three operators for me to interview. They all had received refresher training in the past six months and passed the test given at the end of the training. The first operator I interviewed was extremely knowledgeable of the evacuation procedures; so much so that if an alarm did go off while I was on site I hoped to be within three feet of this individual. The second operator was not quite as knowledgeable but seemed to understand the basics. The third operator proved interesting. He was pretty much clueless, and I was concerned that he would die at his work station should there be a real emergency (a bit of hyperbole here). I also found out that this individual had to take the test three times before he passed. What conclusions should I have drawn from this sample? Were the results material enough for a finding?

In this case, I wasn't sure the rule of three should apply. Maybe the one poor result was significant enough to result in a finding. But still, it was only one individual. How would the site react? I also wondered about the fact that it was the site that selected the three individuals. Why didn't I get three "stars" as is most often the case? Was there something else going on here? A hidden agenda perhaps? The situation became extremely complicated. Ultimately, I decided that the response of the one individual did not meet the materiality test of a finding but I did certainly report the issue to the plant manager and expected him to respond appropriately.

I have presented the above case study to numerous classes during auditor training workshops over the years. It always, and I mean always, results in heated discussions. A class at Duke Energy had a really interesting take on the issue. They mentioned that at some of their coal-fired power plants they hire workers to clean up the coal debris left around the automatic rail car unloading stations. It is a necessary and important job, but they commonly hire local people who are mentally challenged in some way but can do the job just fine. The employees are happy; town is happy; the site is happy; the company meets its diversity goal; a win all around. Yet many of the Duke staff in the class believed that some of these people would fall into the category of the "clueless" person in the case study. And they were concerned that if there was a finding associated with a person not being knowledgeable of emergency response procedures that the person would be identified and fired. This would please no one especially the Human Resources Department. The suggestion of some, which I supported, was to handle the issue outside the structure of the audit program and to look for innovative solutions to the problem, such as a "buddy system." To be fair, others at Duke had a different take. They believed it to be a finding and would include it in the report. To me personally, this is a real tough one and a true dilemma for an auditor.

9. Be Ready for Surprises

If you audit long enough you will experience your fair share of surprises. Any number of auditors on teams of which I have been a member have gotten sick during the audit week. If you're not careful about what you eat and drink, especially in foreign countries, illness can strike fast and hard. I also have had an audit team member stung by a bee, not knowing he was allergic. He had to be taken to the emergency room and treated for a very serious allergic reaction. On another audit of a petroleum refinery, the plant caught fire on early Wednesday morning. When the audit team arrived at the main gate that morning, we were turned away and not allowed to enter the site. The team spent the better part of the day at a local diner awaiting the all clear message. Auditors need to be flexible enough to adjust to these surprises, including my most interesting one, described below.

Angolan Chickens in Uberaba. I was fortunate enough to be a third-party observer of a process safety management audit in Uberaba, Brazil. It was a DuPont Titanium Dioxide plant. During the initial site tour we noticed about two dozen strange birds that had the run of the place (see Figure 13.4). Biggest surprise ever. It turned out that they were Angolan chickens (a.k.a., guinea fowl or guinea hens). When we asked why the chickens were there we were told that they were site pets, kept because they were scorpion predators, which were prevalent in that region of Brazil. Sounded strange to me, but there were certainly no scorpions around. I left the site still wondering about that story. A couple of years later I was teaching an audit course for Schering-Plough and put the Uberaba photograph up on the screen. I asked the question of the 70 attendees: "Why are there Angolan chickens at this plant?" No one, before or since, has known the answer to that question. But in this class, one student yelled out – "scorpion control." It just so happened that this fellow was from Brazil. Question asked, question answered, story confirmed.

10. Treat Others the Way You Want to be Treated

This is an old adage but applies so well to the auditing profession. There is no room for a "good cop, bad cop" scenario in auditing. Balanced professionalism mixed in with a dose of empathy is what is needed. Auditors should above all be:

- Precise
- Thorough
- Reasonable
- Supportive

FIGURE 13.4
Angolan Chickens in Uberaba, Brazil

The bottom line is, try not to be the corporate seagull. That's the auditor who flies to a site, squawks a lot, eats the food, craps all over, and flies home!

Enjoy yourself—every audit is a learning experience!

Notes

1. Cahill, L. B. and L. B. Michelin, "Achieving Quality Environmental Audits: Twenty Tips for Success," *Total Quality Environmental Management*, John Wiley & Sons, Inc., New York, NY.

2. General Permit to Discharge under the Texas Pollutant Discharge Elimination System, TXR050000, Effective August 14, 2011, Part IV, Benchmark Monitoring Requirements, Section A1, Monitoring for Benchmark Parameters in Discharges.

Afterword

Reflections on the First Earth Day—April 22, 1970

"Courage is what it takes to stand up and speak; courage is also what it takes to sit down and listen."

—Winston Churchill

The environmental movement in the United States really began on the first Earth Day on April 22, 1970. It also was helpful in sparking and fostering the concept of environmental auditing just a few years later. The first Earth Day, which I participated in as a college student, was nothing like what the annual April 22 event is in this day and age. What follows are my reflections on that day from an essay I wrote in 2012 for Environmental Resources Management's (ERM) celebrations. It struck an emotional chord with both my peers and the younger staff in the company. I thought it worthwhile to end this auditing story with that essay.

Personal Reflections on the First Earth Day—April 22, 1970

I remember that first Earth Day so well. I was there, sitting on the lawn at Boston Commons, listening to the "rabble-rousers" deliver their speeches, hoping to change the world. They say that some 20 million people attended rallies across the nation on that Wednesday. For me, the experience helped me decide that this environmental "thing" could and should be my vocation. I was a senior in college, a soon-to-be mechanical engineering graduate from Northeastern University with some co-op work experience in environmental engineering. I was on my way to a long and satisfying career.

There are some things that many young people today don't know about that spring of 1970. That school semester was the great national college student strike protesting the then very unpopular Vietnam War. I did not attend one single class but did attend more than my fair share of political rallies. I remember being in support of the Black Panthers taking over the school's administrative building, although, quite honestly, I had no idea who the Black Panthers were. Looking back, I'm sure that many of the college students attending that first Earth Day were probably there on a Wednesday because they were bored and had nothing better to do. After all, they weren't going to class.

My conservative engineering professors were not pleased with our actions and threatened to fail every last one of us. This meant another semester of studies if we wanted to graduate. Northeastern being a co-op university, this also meant taking 5-1/2 years to graduate, not a pleasant prospect for a 22 year old. Finally, a compromise was reached, and we all graduated with a final set of "pass-fail" grades; that is, those of us who were willing to complete a modest special project for each course.

It was a volatile time for the nation and its college students. Each of the males due to graduate in the summer of 1970 had already been through the first Selective Service military lottery, shown on television on December 1, 1969. That night was very emotional for us all. Those with low numbers (based on your month and day of birth) knew they were all but drafted, so many joined the National Guard as an alternative. Those in the middle were the most conflicted. They knew not their fate. Those with high numbers said their prayer of thanks and cried their tears of joy. Thus, we knew long before Earth Day whether we were likely to be drafted upon graduation and that colored how we viewed our futures. My number was 282 out of 366, pretty high, and I was not drafted. Still, I had plenty of classmates who were.

As I look back on it, there were a lot of "firsts" during my senior year: first Earth Day, first national student strike, first draft lottery, and first in my family to graduate from college. After experiencing all these firsts, I then began my career as a noise control specialist for Exxon Research and Engineering, of all things. But that's another story.

History teaches, it's up to us to learn,
Larry Cahill
Exton, PA
April 22, 2012

About the Authors

Author

Lawrence B. Cahill, CPEA

Mr. Cahill is currently a Technical Director in Environmental Resources Management's Performance & Assurance Group located in Malvern, PA. He has over 35 years of professional EH&S experience with industry and consulting. He is the principal author of the widely used text, *Environmental, Health and Safety Audits*, published by Government Institutes, Inc. and now in its 9th Edition. He contributed four chapters in the 1995 text *Auditing for Environmental Quality Leadership*, published by John Wiley & Sons, Inc. Mr. Cahill has published over 60 articles and has been quoted in numerous publications including the *New York Times* and the *Wall Street Journal*. He has been a guest lecturer at the graduate schools of Duke University, Northwestern University, and the New Jersey Institute of Technology.

Mr. Cahill holds a BS in Mechanical Engineering from Northeastern University where he was elected to Pi Tau Sigma, the International Mechanical Engineering Honor Society. He also holds an MS in Environmental Health Engineering from the Robert R. McCormick School of Engineering and Applied Science of Northwestern University, and an MBA from the Wharton School of the University of Pennsylvania where he was awarded a Fels Center of Government Fellowship. He previously was a Project Engineer with Exxon Research and Engineering where he audited petrochemical plants in the U.S., Canada, and Europe. He also was a Principal with the management consulting firm Booz Allen Hamilton. He served as an Environmental Commissioner for the City of Camden, NJ.

Mr. Cahill has been awarded Distinguished Instructor status by ABS Government Institutes. He taught ABS's Environmental, Health and Safety Audits course for over 25 years and is a certified trainer for The Auditing Roundtable. He is a Certified Professional Environmental Auditor with a *Master Certification* and was an original Board Member and former Chairman of the Training and Education Committee of the Board of Environmental, Health and Safety Auditor Certifications (BEAC). He is a long-standing member of The Auditing Roundtable.

Contributing Author

Robert J. Costello, PE, CPEA, Esq.

Robert J. Costello was the co-author with Mr. Cahill on the original articles that were the basis for chapters 4, 7, 8, and 9 in this book. He is a Partner at Environmental Resources Management in Malvern, PA. He has 20 years of professional environmental resource management and consulting experience. Mr. Costello manages global regulatory compliance, management systems, and sustainability assurance programs and typically participates on-site in 30 or more audits and assessments per year. He holds a B.S. in Environmental Engineering from Wilkes University, an M.S. in Environmental Engineering from Syracuse University, and a J.D. from Syracuse University. Mr. Costello is admitted to the bar in Pennsylvania, is a licensed professional engineer in Pennsylvania and Delaware, and is a Certified Professional Environmental Auditor.

Index